Gastrointestinal Endoscopy

Jacques Van Dam, M.D., Ph.D.
Stanford University Medical School
Stanford, California, USA

Richard C.K. Wong, MB., B.S., F.A.C.P.
Case Western Reserve University
Cleveland, Ohio, USA

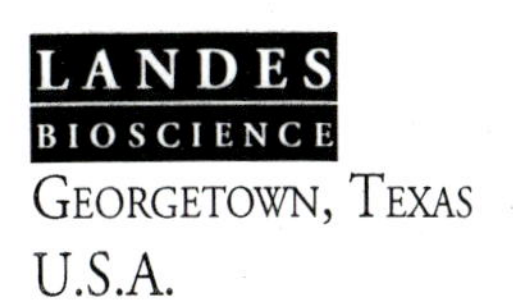
LANDES BIOSCIENCE
GEORGETOWN, TEXAS
U.S.A.

VADEMECUM
Gastrointestinal Endoscopy
LANDES BIOSCIENCE
Georgetown, Texas U.S.A.

Printed in the U.S.A.

Please address all inquiries to the Publisher:
Landes Bioscience, 810 S. Church Street, Georgetown, Texas, U.S.A. 78626
Phone: 512/ 863 7762; FAX: 512/ 863 0081

ISBN: 1-57059-572-0

Library of Congress Cataloging-in-Publication Data

Gastrointestinal endoscopy / [edited by] Jacques Van Dam, Richard C.K. Wong.
p. ; cm. -- (Vademecum)
Includes bibliographical references and index.
ISBN 1-57059-572-0
1. Endoscopy. 2. Gastrointestinal system--Examination. I. Van Dam, Jacques. II. Wong, Richard C. K. III. Series.
[DNLM: 1. Endoscopy, Gastrointestinal--methods. 2. Gastrointestinal Diseases--diagnosis. 3. Gastrointestinal Diseases--therapy. WI 141 G2576 2003]
RC804.E6G372 2003
616.3'307545--dc22

2003024115

Contents

Editors

Jacques Van Dam, M.D., Ph.D.
Stanford University School of Medicine
Stanford, California, USA
Chapter 17

Richard C.K. Wong, MB., B.S., F.A.C.P.
Case Western Reserve University
Cleveland, Ohio, USA
Chapters 15, 17

Contributors

Laurence S. Bailen
Division of Gastroenterology
New England Medical Center
Boston, Massachusetts, U.S.A.
Chapter 14

David Bernstein
Department of Clinical Gastroenterology
Center for Liver, Biliary and Pancreatic Diseases
Winthrop University Hospital
Mineola, New York, U.S.A.
Chapter 25

Manoop S. Bhutani
Department of Medicine
Center for Endoscopic Ultrasound
University of Florida
Gainesville, Florida, U.S.A.
Chapter 22

David J. Bjorkman
Division of Gastroenterology
University of Utah Health Sciences Center
Salt Lake City, Utah, U.S.A.
Chapter 10

William R. Brugge
Gastrointestinal Unit
Massachusetts General Hospital
Boston, Massachusetts, U.S.A.
Chapter 21

David R. Cave
Department of Gastroenterology
St. Elizabeth's Medical Center of Boston
Brighton, Massachusetts, U.S.A.
Chapter 16

Jeffrey S. Cooley
Department of Gastroenterology
St. Elizabeth's Medical Center of Boston
Brighton, Massachusetts, U.S.A.
Chapter 16

Amitabh Chak
Division of Gastroenterology
University Hospitals of Cleveland
Cleveland, Ohio, U.S.A.
Chapter 23

Francis A. Farraye
Harvard Vanguard Health Care
Brigham and Women's Hospital
Boston, Massachusetts, U.S.A.
Chapters 7, 8

Nathan Feldman
Harvard Vanguard Health Care
Brigham and Women's Hospital
Boston, Massachusetts, U.S.A.
Chapters 7, 8

Victor L. Fox
Department of Pediatrics
Children's Hospital
Harvard Medical School
Boston, Massachusetts, U.S.A.
Chapter 26

Martin L. Freeman
Hennepin County Medical Center
University of Minnesota Medical Center
Minneapolis, Minnesota, U.S.A.
Chapter 20

John S. Goff
Department of Medicine
University of Colorado
Denver, Colorado, U.S.A.
Chapter 12

James M. Gordon
Chapter 5

Gerard Isenberg
Cleveland VAMC
Case Western Reserve University
Cleveland, Ohio, U.S.A.
Chapter 19

Brian Jacobson
Boston University Medical Center
Brigham and Women's Hospital
Boston, Massachusetts, U.S.A.
Chapters 7, 8

Philip E. Jaffe
Department of Clinical Gastroenterology
The University of Arizona
University Medical Center
Tucson, Arizona, U.S.A.
Chapter 2

Rome Jutabha
Division of Digestive Diseases
UCLA School of Medicine
Los Angeles, California, U.S.A.
Chapter 11

Kenji Kobayashi
Division of Gastroenterology
University Hospitals of Cleveland
Cleveland, Ohio, U.S.A.
Chapter 23

Shawn Mallery
Division of Gastroenterology
Hennepin County Medical Center
Minneapolis, Minnesota, U.S.A.
Chapter 24

Mark H. Mellow
Digestive Disease Specialists, Inc.
University of Oklahoma School of Medicine
Oklahoma City, Oklahoma, U.S.A.
Chapter 13

Lori B. Olans
Division of Gastroenterology
New England Medical Center
Boston, Massachusetts, U.S.A.
Chapter 14

Peder J. Pedersen
Division of Gastroenterology
University of Utah Health Sciences Center
Salt Lake City, Utah, U.S.A.
Chapter 10

Peter A. Plumeri
School of Osteopathic Medicine
University of Medicine and Dentistry
Sewell, New Jersey, U.S.A.
Chapter 1

Jeffrey L. Ponsky
Department of General Surgery
The Cleveland Clinic Foundation
Cleveland, Ohio, U.S.A.
Chapter 15

Patrick G. Quinn
Northern New Mexico Gastroenterology
University of New Mexico School of Medicine
Santa Fe, New Mexico, U.S.A.
Chapter 9

Douglas K. Rex
Indiana University School of Medicine
Indiana University Hospital
Indianapolis, Indiana, U.S.A.
Chapter 18

Sammy Saab
Division of Digestive Diseases
UCLA School of Medicine
Los Angeles, California, U.S.A.
Chapter 11

Steven J. Shields
Division of Gastroenterology
Brigham and Women's Hospital
Boston, Massachusetts, U.S.A.
Chapter 6

Rosalind U. van Stolk
Hinsdale, Illinois, U.S.A.
Chapter 4

Gregory Zuccaro, Jr.
Section of Gastrointestinal Endoscopy
The Cleveland Clinic Foundation
Cleveland, Ohio, U.S.A.
Chapter 3

CHAPTER 1

Informed Consent

Peter A. Plumeri

Introduction

The process of informed consent is a well established medical practice which has its roots in ethics and law. The requirement to obtain informed consent is based on the ethical principle of self determination and is strengthened in application by both statutory and case law. As a general rule, with few exceptions, and endoscopist is required to obtain informed consent prior to the performance of any endoscopic procedure.

Principles

The principle of elements of informed consent are straight forward. In order to obtain legally adequate informed consent the endoscopist must do the following:[1]

- Disclose the nature of the proposed procedure;
- Its benefits;
- Its risks; and
- Any alternatives.

It must be kept in mind that informed consent is not a form but rather a process of disclosure. This process requires an interaction between the physician and patient.

While a form with appropriate signatures may evidence that the process took place, standing alone it is not to be equated with properly executed informed consent.

Adjunctive aids such as video tapes[2] and informational brochures aid in the disclosure process and enhance patient understanding and are to be commended. These aids do not on their own act as a substitute for the physician patient interaction.

Method

The endoscopist should interact directly with the patient at some point prior to the endoscopy to meet the requirements of informed consent.

The description of the nature of the procedure should include the methodology of the process including planned sedation.

Risks need to be disclosed. Not every possible risk needs to be reviewed but those which occur with greater frequency and those of a serious nature should be included. For endoscopic procedures in general the risk of perforation should be outline and for endoscopic retrograde cholangiopancreatography the risk of pancreatitis should be defined. In addition, the need for surgical intervention in the event of a realized complication needs disclosure.

The patient should understand the benefits of the proposed procedure and these should be defined. Alternatives, even those more hazardous than the procedure,

Gastrointestinal Endoscopy, edited by Jacques Van Dam and Richard C. K. Wong.

need elucidation. For example, disclosure of surgical alternatives to achalasia dilation and colonic polypectomy should be reviewed.

It is generally a sensible practice to have the process of informed consent witnessed if possible. Further, the interaction of informed consent should be documented in the record.

Exceptions

It may not always be possible to obtain informed consent and there are several exceptions. They include:

- **Emergency**
  In the event of a threat to life (e.g., massive variceal bleeding) this exception can be applied. In using this exception documentation of the urgency of the situation is needed.
- **Incompetency**
  An incompetent patient cannot give adequate informed consent. In cases such as these the endoscopist must seek out the next responsible party or the patient and obtain informed consent.
- **Waiver**
  The patient may waive the disclosure process. Waiver is a valid exception provided it is knowing and voluntary on the part of the patient.
- **Therapeutic privilege**
  This rare exception can be used if the physician believes the disclosure would so damage the decision making of the patient that the patient would be unduly harmed. In applying this exception the endoscopist should seek out the support of another professional or family member and document the process.
- **Legal mandate**
  An endoscopist may carry out a procedure on a noncompliant patient if ordered by a court of law or by statute.

Selected References

1. American Society of Gastrointestinal Endoscopy, Risk Management Handbook, 1990.
2. Agre P, Kurtz R, Krauss B. A randomized trial using videotape to present consent information for colonoscopy. Gastrointest Endosc 1994; 40:271-276.

CHAPTER 2

Conscious Sedation and Monitoring

Philip E. Jaffe

Introduction

- The term "conscious sedation" has been used to describe the use of mood altering, amnestic, and analgesic medications before and during procedures to improve patient tolerance and satisfaction. In the United States this technique has seen widespread use in the field of gastrointestinal endoscopy and has become standard practice in most areas. Interestingly, this is not the case when the same procedures are performed in many European, Asian, South American, and African countries so the need for the use of sedating medications may vary according to regional differences in patient expectations, standard medical practices, and societal norms. In the United States, Canada, and the United Kingdom however, most patients prefer some sedation when undergoing upper endoscopy, colonoscopy, or ERCP.
- Recently, a special task force of American Society of Anesthesiologists met to review publications and expert opinions on the appropriate use of sedation and analgesia by nonanesthesiologists and created several publications on guideline and suggestions based on an evidence-based approach to this topic. These guidelines were endorsed by the American Society for Gastrointestinal Endoscopy and have been incorporated into appropriate sections of this chapter. The imprecise term "conscious sedation" has been replaced by "sedation and analgesia" and has been defined as "a state that allows patients to tolerate unpleasant procedures while maintaining adequate cardiorespiratory function and the ability to respond purposefully to verbal command and/or tactile stimulation". The purpose of this definition is to more unambiguously describe the nature and goals of this type of procedural sedation and to allow for more universally acceptable standard of practice.
- There are a number of reasons to use sedation and analgesia with gastrointestinal endoscopic procedures. These include the reduction of anxiety and pain, the induction of amnesia, and improvement in patients cooperation. The type and amount of sedation and analgesia used will depend of characteristics of the procedure to be performed (e.g., length and amount of discomfort or anxiety provoked), individual patient factors (e.g., age, underlying medical problems, level of anxiety, prior experience with endoscopic procedures, and current use of anxiolytic or opiate medications), patient preferences, need for repeated procedures in the future and the degree of patient cooperation needed during the procedure. The details of how these factors interact and specific recommendations will be discussed in the "technique" section.

Gastrointestinal Endoscopy, edited by Jacques Van Dam and Richard C. K. Wong.

- Close monitoring of patients undergoing endoscopic procedures with sedation is essential because of the risk of cardiopulmonary complications related to the use of these medications. The most important monitors are not automated blood pressure or pulse oximetry machines but trained ancillary personnel who can detect signs of respiratory depression and allow for prompt intervention. While continuous ECG machines, pulse oximeters, and automated blood pressure and respiratory monitors can be helpful in many situations, they are no replacement for a well-trained and attentive assistant. Because the risk of serious or life-threatening complications related to the use of sedating or analgesic medications with endoscopic procedures is very small (e.g., risk of death <1:10,000), it would take studies with enormous numbers of patients to prove that any monitoring intervention will improve the overall safety of the procedure. Therefore, recommendations on pre, intra-, and post-procedure monitoring with endoscopic procedures are based on an understanding of the nature of the procedures and pharmacodynamic properties of the medications and not the results of well-designed prospective studies.

Indications and Contraindications

- Preprocedural assessment
  - In any patient where the use of sedation or analgesia is being considered, a careful and directed history and physical examination should be performed. The main purpose is to identify factors that may be useful in optimizing the safety and effectiveness of the procedure. As with the indications and contraindications for the performance of any procedure, the potential risks of using sedating medications must be weighed against the expected benefits.
  - **Preprocedural history** should include preexisting medical conditions including cardiopulmonary disease, renal disease, psychiatric disorders, and prior surgery. Drug allergies and intolerances as well as experience with prior exposure to sedating or analgesic medications should be asked about. Current medication use is important as some medications can interact with sedating medications, and the concurrent use of opiates or benzodiazepines may necessitate different strategies for achieving adequate sedation. Alcohol, tobacco, and substance abuse history should be elicited. Patients with prior adverse reactions to specific sedating medications should be given alternatives, and those who have had prior intolerance to traditional sedation/analgesia regimens should be considered for general anesthesia depending on the indication for the procedure. Patients who are using alcohol or illicit substances at the time of the procedure may be difficult to sedate and pose a high risk for complications, and therefore one should have a low threshold for requesting the assistance of an anesthesiologist to achieve adequate sedation. In the setting of truly elective procedures, these patients should be rescheduled for a time when they are not taking these substances if possible.
  - The focused **physical examination** should concentrate on vital signs, airway, heart, and pulmonary status. Patients determined to have new, clinically significant heart murmurs, cardiac arrhythmias, signs of heart failure, or wheezing should not undergo elective procedures with sedation/analgesia until they have been formally evaluated and their clinical condition optimized.

Table 2.1. ASA classification system for physical status

ASA Class	Disease State
Class I	No organic, physiologic, biochemical, or psychiatric disturbance
Class II	Mild to moderate systemic disturbance
Class III	Severe systemic disturbance
Class IV	Severe, life-threatening systemic disturbance
Class V	Moribund patient who has little chance of survival

- The **patient's overall risk** can be summarized using the American Society for Anesthesiologists classification scheme for medical states (Table 2.1). This can be easily calculated prior to any procedure and may be useful in determining the type and amount of sedation to use.
- Prior to undergoing any endoscopic procedure it is helpful to discuss the specific nature of the procedure and the use of sedating medications to alleviate anxiety and discomfort. This discussion can be very helpful in reducing patient anxiety that is a major source of procedural intolerance. Person-to-person discussion or videotaped narratives have been used successfully toward this end.

- Specific procedures where sedation/analgesia may be indicated
  - **Upper endoscopy**. Although esophagogastroduodenoscopy (EGD) is a relatively short procedure generally lasting less than 10 minutes it can be the source of great anxiety and some discomfort because of the potential for gagging, retching, and discomfort due to distention of the upper gut with air. When using larger (therapeutic endoscopes) there is a greater need for sedation due to the increased discomfort associated with them. However many patients can get by with little or no medication when narrow caliber instruments are used in association with topical pharyngeal anesthetics. Sedation with benzodiazepine with or without small amounts of opiates is well suited to this situation. When percutaneous gastrostomy (PEG), dilation, electrocautery, or laser therapy is performed in conjunction with EGD greater doses of opiates are used because of the added pain associated with these ancillary procedures.
  - **Colonoscopy**. Patient undergoing colonoscopy nearly always prefer some form of sedation / analgesia due to the discomfort causes by distention of the colon with air and stretching of the large bowel as the endoscope is advanced beyond the left colon. There has been some recent experience performing unsedated colonoscopy revealing that many patients (particularly U.S. veterans) can tolerate this, however these studies have not yet been expanded into the nonveteran population.
  - **ERCP**. Because of the prolonged time required to perform this procedure, the associated discomfort, and the need for greater patient cooperation nearly all patients undergoing this procedure require significant sedation. Some centers have begun experimenting with patient controlled anesthesia (PCA) delivery systems to minimize the doses of medications used and improve patient satisfaction.

  - **Endoscopic ultrasound (EUS)**. EUS of the upper GI tract usually requires sedation due to the fact that it generally takes much longer to complete than diagnostic upper endoscopy and requires greater patient cooperation. This is particularly true when retroperitoneal assessment of the pancreas, biliary tree, and associated vascular structures is needed. EUS of the rectum and anus can usually be done without sedation unless there is a condition that will likely lead to discomfort with the procedure (e.g., perirectal abscess).
- Procedures not requiring sedation
  Although some gastrointestinal procedures may required sedation/analgesia in selected circumstances (e.g., an extremely anxious patient with low pain threshold), they can be easily performed without systemic medication under most conditions. These would include flexible sigmoidoscopy, rigid sigmoidoscopy, anoscopy, percutaneous liver biopsy, gastrostomy removal using traction, peroral esophageal dilation, and paracentesis.

Equipment and Accessories

In any GI lab where sedation/analgesia is to be performed, certain devices and accessories need to be available for routine use during cases and for emergency use in potentially life threatening situations.

- Monitoring equipment
  - **Automated blood pressure and heart rate monitoring equipment** should be used before, during, and after sedated procedures and have the capacity to provide a hard copy display of recordings. While manual equipment can be used it would require excessive personnel time for continually measuring these parameters and recording them.
  - **Pulmonary ventilation** should be measured by auscultation of breath sound and observing chest movement. The latter can be facilitated by using automated devices frequently included with multi-functioning vital sign recording systems.
  - **Pulse oximetry** using finger or ear probe sensing devices and utilizing a visual and audio display has become increasingly popular for monitoring during endoscopic procedures. While fairly accurate in measuring oxygen saturation, this measure can be falsely reassuring as oxygenation can be well maintained early in the course of hypoventilation while hypercarbia develops.
  - Continuous **ECG monitoring** should be available for patients and/or procedures at increased risk for cardia dysrhythmias. In many centers these are used routinely for all sedated procedures.
- Ancillary equipment / Personnel
  - Intravenous supplies and a variety of sterile infusion solutions, supplemental oxygen, oral suctioning equipment, nasal cannulae, facemasks, and oral airways should all be available within the GI lab. Resuscitation equipment including an Ambu bag, defibrillator, emergency medications (such as epinephrine, atropine, and lidocaine), and intubation equipment should be readily accessible in the event of an emergency.
  - GI assistants and other medical personnel within the GI lab should be trained and certified in Basic Life Support and at least one individual involved with each sedated procedure should be trained and certified in Advance Life Support measures. In addition there should be an ongoing educational effort in instructing GI lab staff on the appropriate use of the various medications used routinely in sedation/analgesia.

- Medications

  Medications used in sedation/analgesia for GI procedures can be generally categorized as benzodiazepines, opiates, neuroleptics, or sedative hypnotics. These agents are frequently combined to achieve the desired endpoint of anxiolysis, amnesia, analgesia, and cooperation. The specific use of these agents will depend on patient characteristics as well as the nature of the procedure to be performed and will be discussed in more detail in the "technique" section. A summary of the pharmacodynamic properties of these medications is included (Table 2.2).

  - **Benzodiazepines.** The most commonly used agents in this category include midazolam and diazepam. These drugs act centrally to induce a dose-related state of sedation, anxiolysis, muscle relaxation, and amnesia. The most common side effects are diminished ventilation, mild hypotension, and pain and phlebitis with the injection of diazepam. The latter problem can be overcome with the use of the lipid emulsion of diazepam (Diazemuls R).
  - **Opiates.** These drugs act centrally on the ??-opioid receptor to induce a state of euphoria and analgesia. Respiratory depression can be seen at relatively low doses due to these drugs' depression of hypoxic and hypercarbic respiratory drive. The most commonly used opiates in GI endoscopy include meperidine, morphine sulfate, and fentanyl. While there are theoretical advantage to the use of fentanyl and other newer narcotic agents (more rapid onset of action and shorter half-life), meperidine continues to perform well in clinical practice. When opiates are used in combination with benzodiazepines they exhibit significant synergy which can increase the additive effect of these drugs up to eight-fold.
  - **Neuroleptics.** Droperidol is an anti-nausea medication structurally related to haloperidol frequently used in post-anesthesia recovery units prior to endotracheal extubation. When used in combination with benzodiazepines and opiates it produces a state called "neuroleptanalgesia" which is characterized by quiescence, reduced motor activity, and indifference to pain and other stimuli. It is particularly useful in patients who require high doses of benzodiazepines and opiates and are unable to cooperate. The major side effects are hypotension and prolonged recovery.
  - **Sedative hypnotics.** Propofol is a relatively new agent used for induction and maintenance of sedation or anesthesia. It has been applied to use with GI endoscopic procedures because of its very rapid onset of action and short half-life. It has the potential for severe ventilatory depression (dose related) and is restricted for use by anesthesiologists in most centers.
  - **Reversal agents.** Flumazenil is the only available benzodiazepine receptor antagonist available and has a high affinity for this receptor essentially totally reversing the central effects of these agents. It should be recognized that the biological half-life of this drug is significantly shorter than that of the biologically active metabolites of most benzodiazepine so there is a theoretical risk of late resedation. In addition, this drug can precipitate benzodiazepine withdrawal reactions including seizures in patients chronically using these drugs so a careful history should be obtained before giving flumazenil. Fortunately, neither resedation nor benzodiazepine withdrawal is seen frequently in clinical practice. Naloxone is a useful opiate antagonist that like

Table 2.2. Pharmacologic properties of commonly used medications for sedation/analgesia

Drug	Onset	Peak Effect	Duration of Action
Midazolam	Immediate	1-5 min	1-2 hrs
Diazepam	3-5 min	5 min	15-60 min
Meperidine	1 min	5-7 min	2-4 hrs
Fentanyl	1-2 min	5-15 min	30-60 min
Morphine sulfate	1-5 min	20 min	4-5 hrs
Droperidol	3-10 min	15-30 min	3-6 hrs
Propofol	40 sec	1 min	5-10 min
Naloxone	2 min	5-15 min	45 min
Flumazenil	1-5 min	6-10 min	2-4 hrs

flumazenil's reaction in chronic benzodiazepine users can also precipitate severe withdrawal reactions when given to individuals chronically using narcotics. Its action is prompt, specific to opiates, and has a shorter effect than the half-life of most opiates.

Technique

- Preparation
  - There should be a **standardized form** used to record all facets of the pre-, intra-, and post-procedure data and events.
  - All patients should undergo a **preprocedural assessment** including baseline vital signs, temperature, and directed history and physical exam (see indications and contraindications).
  - It should be verified that the patient has **fasted** appropriately (at least 6-8 hours for solids and 2-3 hours for clear liquids) and has an appropriate **chaperone** for post-procedure discharge.
  - **Intravenous access** with an indwelling catheter should be secured before any GI procedure requiring sedation/analgesia.
  - Oxygen saturation using **pulse oximetry** should be measured before and then continuously during the procedure.
  - **Supplemental oxygen** should be considered in individuals felt to be a high risk for desaturation during the procedure due to underlying cardiopulmonary disease, a complex or lengthy anticipated procedure, or preprocedure desaturation.
  - Similarly continuous **ECG** monitoring should be considered in elderly patients or those with a history of cardiac dysrhythmias or dysfunction.
  - Heart rate, blood pressure, and respiratory rate should be measured before the procedure, every 2-5 minutes during, and at least every 5 minutes during recovery following the procedure.
  - When available, data on the amount, type and reaction to sedating medications used with previous procedures should be noted to give a "ball park" figure for what will likely be required for the current procedure.
- Specific medications
  - **Benzodiazepines**: Midazolam given at 0.5-2 mg. intravenous boluses every 2-5 minutes or diazepam 1-4 mg every 2-5 minutes.

- **Opiates**: Meperidine 25 mg intravenous boluses every 3-5 minutes or fentanyl 25-50 μg every 3-5 min or morphine sulfate 1-4 mg every 5-10 min.
- **Neuroleptics**: Droperidol 1-2.5 mg intravenous boluses every 8-10 min.

- Specific procedures
  - **EGD**. Benzodiazepines alone (generally 2-5 mg of midazolam or 5-10 mg of diazepam) or in combination with relatively low doses of opiates (25-50 mg of meperidine, 25-75 μg of fentanyl, or 2-5 mg of morphine sulfate) are required to achieve adequate sedation for most diagnostic EGD's. Benzodiazepines are given predominantly since the main goal is anxiolysis and not analgesia. The use of pharyngeal anesthetic agents such as lidocaine or benzocaine may be helpful in reducing discomfort when larger instruments are used or low doses of (or no) intravenous sedation is given.
  - **Colonoscopy**. In this procedure benzodiazepines are used in combination with higher doses of opiates since there is a greater potential for pain related to this examination. It is important to remember that opiates and benzodiazepines act in synergy to produce the desired level of sedation and analgesia but also may potentiate side effects such as hypoventilation and hypotension. Typical doses of medications for this procedure would be 2-7 mg of midazolam or 8-15 mg of diazepam in combination with 50-100 mg of meperidine, 50-150 _g of fentanyl, or 5-10 mg of morphine sulfate.
  - **ERCP and EUS**. Since these procedures usually last longer and require a greater degree of patient cooperation, it is not unusual for higher doses of benzodiazepines and opiates to be required. In addition, droperidol may be well suited for these procedures due to these same factors. Typical doses of medications for these procedures might be 5-10 mg of midazolam or 10-20 mg of diazepam in combination with 75-150 mg of meperidine, 100-250 μg of fentanyl, or 10-15 mg of morphine sulfate as well as 2.5-5 mg of droperidol.
- Recovery/discharge. Monitoring must be continued during the recovery phase following procedures because of the prolonged biological activity of most agents used with sedation and the loss of stimulation (usually caused by the endoscope and assistant) upon completion of the examination. Periodic measurement of heart rate, blood pressure, pulse oximetry, and respiratory rate should continue at at least 5 min intervals during recovery. It is helpful to have formal discharge criteria to determine when patients are appropriate to be discharged to home following any procedure require sedation/analgesia. The exact time required for recovery prior to discharge will vary greatly depending of the age and medical condition of the patient and the type and amount of sedation used.

Outcome

The desired goal of sedation and analgesia with GI endoscopy procedures is to improve the overall patient tolerance of procedures by reducing anxiety and pain while maintaining the highest possible margin of safety. By understanding the actions, interactions, and pharmocokinetics of medications used for sedation and analgesia this can be accomplished in the vast majority of patients. In addition, most complications are readily treatable or reversible if recognized early so vigilance in monitoring is important in ensuring a safe outcome.

2

Complications

- **Cardiopulmonary complications** account for the greatest number of serious side effects related to the use of sedation and analgesia with endoscopic procedures. They are the most common complication of any sort associated with most GI endoscopic examinations. When hypotension and oxygen desaturation are included the rate of cardiopulmonary side effects is 0.2-0.5%. The death rate directly related to sedation is less than 1:10,000 but remains the most likely cause of death related to performing endoscopic procedures. It is difficult to know if the use of more complex and expensive monitoring equipment can reduce this extremely small risk or improve the overall outcome.
- Other more common adverse events include the **"paradoxical reaction"** to the administration of sedating medications. The most common scenario is the young alcoholic or substance-abusing patient who becomes agitated and difficult to restrain after being given what are usually adequate doses of benzodiazepines and opiates. Giving additional doses of benzodiazepines usually worsens this problem and occasionally the patient may cooperate when droperidol is added. Frequently, however, these patients will require the assistance of an anesthesiologist to administer propofol for better control.
- Another rare but noteworthy complication is related to the use of **topical pharyngeal** anesthetics. Some patients can development of profound oxygen desaturation and cyanosis caused by methemoglobinemia brought on by the use of lidocaine or benzocaine sprays. The treatment is high flow oxygen administration and possibly the use of IV methylene blue.

Selected References

1. The American Society for Gastrointestinal Endoscopy. Sedation and monitoring of patients undergoing gastrointestinal endoscopic procedures. Gastrointes Endosc 1995; 42:626-629.
2. The American Society for Gastrointestinal Endoscopy. Preparation of patients for gastrointestinal endoscopy: Guidelines for clinical application. Gastrointes Endosc 1988; 34(Suppl):32S-34S.
3. Monitoring Equipment for Endoscopy: ASGE technology assessment status evaluation. The American Society for Gastrointestinal Endoscopy 1994.
4. Practice guidelines for sedation and analgesia by nonanesthesiologists: A report by the American society of Anesthesiologists task force on sedation and analgesia by nonanesthesiologists. Anesthesiology 1996; 84:459-471.
5. Complications of gastrointestinal endoscopy. Gastrointestinal Endoscopy Clinics of North America 1996; 6:34-42.
6. Chuah S, Crowson C, Dronfield M. Topical anaesthesia in upper endoscopy. British Medical Journal 1991; 303:695-697.
7. Wilcox C, Forsmark C, Cello J. Utility of droperidol for conscious sedation in gastrointestinal endoscopic procedures. Gastrointes Endosc 1990; 36:112-115.
8. Jaffe P, Fennerty M, Sampliner R et al. Preventing hypoxemia during colonoscopy: A randomized controlled trial of supplemental oxygen. J Clin Gastroenterol 1992; 14(2):114-118.
9. Kost M. Conscious sedation medication pharmacologic profile. In: Kost M, ed. Manual of Conscious Sedation. Philadelphia: W. B. Saunders Co., 1998:266-287.
10. Brown C, Levy S, Susann P. Methemoglobinemia: Life-threatening complication of endoscopy premedication. Am J Gastroenterol 1994; 89:1108-1111.

CHAPTER 3

Antibiotic Prophylaxis

Gregory Zuccaro, Jr.

Introduction

The risk of serious infection attributable to gastrointestinal endoscopy is low. Nevertheless, there are circumstances where antibiotics may be reasonably provided in an attempt to decrease the likelihood of serious infection as a result of an endoscopic procedure. In each case, the risk/benefit ratio of providing antibiotics should be considered. There are potential side effects from antibiotics, including allergic reactions, antibiotic-associated colitis, etc. Indiscriminate use of antibiotics also encourages the development of drug resistant organisms.

There are several scenarios where an infectious complication of endoscopy may occur. Infection may occur remote from the gastrointestinal tract, due to translocation of oropharyngeal or gastrointestinal flora into the bloodstream, with subsequent seeding of a susceptible site. Infection may occur after endoscopic manipulation of an obstructed or diseased region of the gastrointestinal tract, where bacterial colonization may have previously taken place. A potential pathogen may be present on an improperly reprocessed endoscope or accessory and introduced into a susceptible portion of the gastrointestinal tract.

Infection in Areas Remote from the Gastrointestinal Tract

- Infective endocarditis. In order to develop infective endocarditis from an endoscopic procedure, a patient with a cardiac structural abnormality (most commonly a cardiac valvular abnormality) must develop bacteremia with an organism likely to cause endocarditis. There are very few documented cases of infective endocarditis due to endoscopy, despite millions of annual procedures. It is extremely unlikely that a prospective trial proving the efficacy of antibiotic prophylaxis for the prevention of infective endocarditis due to gastrointestinal endoscopy will ever be performed. Nevertheless, it is reasonable under specific circumstances to provide prophylaxis.
  - The cardiac conditions. Cardiac conditions may be stratified into those at highest risk for the development of infective endocarditis, those at moderate risk, and those at lowest risk. These conditions are listed in Table 3.1. The highest risk lesions are typically apparent to the endoscopist by history alone. Diagnosis of the moderate risk lesions, particularly mitral valve prolapse, is more problematic. Many patients presenting to the endoscopy suite will have previously been told of a history of **mitral valve prolapse**. Only some of these patients will actually have this condition, and fewer still will have the particular structural status predisposing them to infective endocarditis. The presence of a cardiac murmur on examination does not necessarily indicate a predisposition for infective endocarditis, nor does the absence of

Gastrointestinal Endoscopy, edited by Jacques Van Dam and Richard C. K. Wong.

3

Table 3.1. Relative risk of cardiac and other conditions for development of infectious complications due to gastrointestinal procedures (adapted from references 5,6)

Highest Risk	Moderate Risk	Lowest Risk
Cardiac Conditions		
• Prosthetic valve	• Mitral valve prolapse	• Coronary artery bypass
• History of endocarditis with insufficiency	• Pacemaker	
• Systemic-pulmonary shunt	• Hypertrophic cardiomyopathy	• Implantable defibrillator
• Most congenital malformations	• Rheumatic valvular dysfunction	
Noncardiac Conditions		
• Synthetic vascular graft (less than 1 year old)	• Synthetic vascular graft (greater than 1 year old)	• Prosthetic joint or orthopedic prosthesis

a murmur on auscultation completely exclude this predisposition. The only way to definitively determine this risk is with echocardiography. In the absence of this information, it is reasonable for the endoscopist to assume the patient has mitral valve prolapse, and treat the patient as if they may indeed have the predisposing lesion, providing prophylaxis for the indicated procedures. If a gastrointestinal condition is discovered where repeat endoscopy at some future time is likely (e.g., peptic stricture), a formal echocardiogram can be recommended before the time of the next endoscopy. Those patients with the lowest risk cardiac conditions are no more likely to experience an infectious cardiac complication from gastrointestinal endoscopy than the general population.

- **Bacteremia** associated with gastrointestinal procedures. The risk of bacteremia associated with gastrointestinal endoscopy is low. The bacteremia rates, as well as the type of organisms encountered, do vary with the procedure performed.
  - **Upper GI endoscopy**. The bacteremia rate associated with upper gastrointestinal endoscopy is 0-4%. The bacteremia is transient, lasting in most cases only a few minutes. Organisms are those found commonly in the oropharynx. Of these organisms, the one most likely to be associated with infective endocarditis is viridans streptococcus. The bacteremia rate does not appear to be significantly altered by mucosal biopsy.
  - **Esophageal stricture dilation**. In contrast to upper endoscopy, approximately 20% of patients undergoing esophageal stricture dilation will experience viridans streptococcal bacteremia. This may be more prolonged, lasting at least 20 minutes in 5% of bacteremic patients.[1]
  - **Variceal sclerotherapy**. This procedure is also associated with relatively higher rates of transient bacteremia as compared to upper endoscopy, with reported values as high as 45%.[2] However, there are several factors contributing to this overall higher rate. Patients with advanced liver disease,

particularly when presenting with acute gastrointestinal bleeding, have a greater than 10% bacteremia rate *before* the procedure is even performed.[3] Further, in some cases bacteremia may be due to nonoropharyngeal organisms such as *Pseudomonas aeruginosa* or *Serratia marcescens*, implicating contaminated endoscopes, water bottles or accessories as the source of the bacteremia. Bacteremia is also more likely when longer length sclerotherapy needles are utilized, or during emergency procedures where greater volumes of sclerosant are typically administered. Interestingly, while there are less data available compared to sclerotherapy, ligation of esophageal varices (banding) is associated with bacteremia rates closer to that of diagnostic upper endoscopy than to injection sclerosis.

- **Sigmoidoscopy/Colonoscopy**. The overall rate of bacteremia attributable to sigmoidoscopy/colonoscopy is similar to that of upper endoscopy. Organisms are most frequently gram negatives unlikely to cause infective endocarditis. However, enterococcal bacteremia is a possibility, and there has been at least one case report of enterococcal endocarditis after flexible sigmoidoscopy.[4] Bacteremia does not increase substantially after biopsy or polypectomy.
- **Endoscopic retrograde cholangiopancreatography**. The documented rate of bacteremia for all patients undergoing this procedure is approximately 5%. In most cases where there is a possibility of biliary obstruction, antibiotics typically are provided more for prevention of cholangitis than for prevention of infective endocarditis.

- Strategies for antibiotic prophylaxis. In general, current recommendations for antibiotic prophylaxis for prevention of infective endocarditis due to gastrointestinal endoscopy are based on the frequency of bacteremia resultant from a given procedure, and the relative risk of the cardiac lesion for the development of infective endocarditis.[5,6] It is recommended that those patients with the higher risk lesions, undergoing procedures associated with the higher bacteremia rates, receive prophylaxis. Conversely, those patients with lower risk cardiac lesions undergoing procedures associated with low bacteremia rates do not require prophylaxis. The limitations in this strategy must be acknowledged. There is no established correlation between transient bacteremia per se as a result of endoscopy and the risk of infective endocarditis. Transient bacteremia occurs with activities of daily living including mastication and defecation for which prophylaxis is typically not provided. Further, most cases of infective endocarditis are *not* due to an intervention such as endoscopy. These facts have prompted some experts to conclude that prophylaxis against infective endocarditis is never necessary in association with gastrointestinal endoscopy.[7]

  I favor the use of prophylactic antibiotics in appropriate circumstances. Infective endocarditis is a condition associated with high morbidity and mortality. If endocarditis develops in a patient several weeks or even months after undergoing an endoscopic procedure, it is conceivable that the infection will be attributed to the procedure, even if it was not in fact the precipitating event. What should be resisted is the inappropriate use of antibiotics, particularly for patients with moderate risk cardiac lesions (e.g., mitral valve prolapse) undergoing procedures associated with low bacteremia rates (e.g., colonoscopy).

- **Prosthetic joints and orthopedic prostheses.** These infections infrequently involve organisms originating in the oropharynx or gastrointestinal tract. There are no data indicating a risk of infection of this nature occurring as a result of gastrointestinal endoscopy. With respect to gastrointestinal endoscopy, these are low risk conditions. Therefore, I do not recommend antibiotic prophylaxis for patients with prosthetic joints or orthopedic prostheses undergoing endoscopy.
- **Synthetic vascular graft.** There are data from animal models suggesting injection of high bacterial titers can lead to infection of a newly placed vascular graft. Once the graft has epithelialized, high titer bacteremia is far less likely to result in infection.[8] For the purposes of a strategy for antibiotic prophylaxis, it is reasonable to consider the patient with a synthetic vascular graft less than one year old at high risk for an infectious complication. Once the graft is greater than one year old, where complete pseudointimal coverage has likely occurred, it can be considered a low risk condition.
- Recommendations for antibiotic prophylaxis. A summary of recommendations for prevention of infectious complications of sites remote from the gastrointestinal tract is presented in Table 3.2. These are consistent with practice guidelines recently published by the American Society for Gastrointestinal Endoscopy.[5] Any situation where the recommendation differs from these societal guidelines is indicated as a footnote to the Table.

Infection in the Area of Endoscopic Manipulation

- Cholangitis after retrograde cholangiography. Patients may present with pain, jaundice, fever and/or leukocytosis due to symptomatic obstruction of the biliary tree. In this setting, it is unlikely that consideration for antibiotic prophylaxis is necessary, as the patient will likely have received antibiotics prior to their arrival in the endoscopy suite. Therefore, this text considers antibiotic prophylaxis for that patient with no overt symptoms or signs of cholangitis.
  There have been several randomized, prospective trials evaluating the efficacy of preprocedural antibiotics for the prevention of cholangitis. Results are mixed, with some studies indicating a benefit,[9,10] but others not.[11-13] Risk factors for the development of cholangitis:
  - **Obstruction** of the biliary tree/inability to relieve obstruction. Partial obstruction of the biliary tree may occur due to tumor, stricture or choledocholithiasis. Cholangitis is less likely as a complication of retrograde cholangiography if injected contrast drains freely from the biliary tree after opacification, as when stones are successfully extracted with papillotomy and extraction, or an obstruction relieved with dilation and/or stenting. Transient bacteremia, and even cholangitis, may still occur if these drainage maneuvers are accomplished, but it appears to be less likely if drainage is established.
  - **Multiple manipulations** of an obstructed system. The patient undergoing sequential manipulations of the obstructed biliary tree is more likely to experience cholangitis.[14] This correlates with difficulty in establishing definitive drainage.
  - **Inappropriate reprocessing techniques.** Contaminated endoscopes, water bottles and other accessories are a potential source of cholangitis, frequently involving gram negatives such as *P. aeruginosa* and *S. marcescens*. This is discussed in more detail in Chapter 4.

Table 3.2. Recommendations for antibiotic prophylaxis for prevention of infection due to gastrointestinal endoscopy involving sites remote from the GI tract

Procedure	Patient Condition	Prophylaxis Recommended	An Acceptable Regimen+
Upper endoscopy Variceal banding	Highest risk	Yes*	Ampicillin 2 gm IV++
	Moderate risk	No	
	Lowest risk	No	
Sigmoidoscopy Colonoscopy Diagnostic ERCP	Highest risk	Yes*	Ampicillin 2 gm IV++ plus gentamicin 1.5 mg/kg (up to 80 mg) IV
	Moderate risk	No	
	Lowest risk	No	
Esophageal stricture dilation Variceal sclerotherapy	Highest risk	Yes	Ampicillin 2 gm IV++
	Moderate risk	Yes*	Ampicillin 2 gm IV++
	Lowest risk	No	

+ Antibiotics may be administered within 30 minutes of procedure start time. *American Society for Gastrointestinal Endoscopy leaves decision to endoscopist discretion. ++Vancomycin 1 gm IV may be substituted for the penicillin allergic patient.

- Organisms encountered. A wide variety of organisms have been cultured from the blood and bile of patients with biliary obstruction. Most frequent are gram negative organisms, but enterococcus and some anaerobes may also be encountered. Polymicrobial infections are not uncommon.[15]
- Recommendations. For the immunocompetent patient with suspected biliary obstruction, in the complete absence of symptoms/signs of acute cholangitis, there is no clear consensus in the literature as to the role of prophylactic antibiotics. For the immunocompromised patient, prophylaxis would appear prudent. If prophylaxis is administered, a broad spectrum antibiotic such as piperacillin 4 grams IV is reasonable. If there is evidence of previous biliary manipulation with incomplete drainage of an obstruction, anti-pseudomonal coverage should be provided (e.g., addition of an appropriate aminoglycoside to piperacillin). If after biliary endoscopy complete drainage and relief of obstruction is not established, broad spectrum antibiotic coverage should be instituted or continued.

- **Injection of a pancreatic pseudocyst.** When retrograde pancreatography results in injection of a pancreatic pseudocyst, there is a possibility of introduction of infection. There are no data to indicate that the provision of prophylactic antibiotics will result in prevention of this occurrence or limitation of morbidity/mortality. The most important aspect of infection prophylaxis here is to plan prior to the opacification of the cyst to provide definitive drainage (whether endoscopic, radiologic or surgical) after the procedure.
- Endoscopic placement of a percutaneous feeding tube. There are data suggesting that antibiotic prophylaxis with cefazolin 1 gram IV decreases the rate of soft tissue infection after feeding tube placement.[16]

Special Cases

- The patient with cirrhosis and ascites. There are few data indicating the risk of infection of ascitic fluid as a result of gastrointestinal endoscopy. For procedures associated with high bacteremia rates (e.g., variceal sclerosis in a patient with portal hypertension and ascites), it is reasonable to provide prophylaxis (see Table 3.2). For procedures where bacteremia rates are low, prophylaxis is not necessary.
- The **immunocompromised patient**. Prophylaxis is reasonable when there is a risk of bacteremia with organisms that may be associated with sepsis, as when retrograde cholangiography is performed in the setting of possible or known obstruction. For procedures where bacteremia rates are low, and organisms unlikely to cause a systemic sepsis, prophylaxis may not be of benefit.

Selected References

1. Zuccaro, G, Richter JE, Rice TW et al. Viridans streptococcal bacteremia after esophageal stricture dilation. Gastrointest Endosc 1998; 48:568-573.
2. Botoman VA, Surawicz CM. Bacteremia with gastrointestinal endoscopic procedures. Gastrointest Endosc 1986; 32:342-346.
3. Ho H, Zuckerman M, Wassem C. A prospective controlled study of the risk of bacteremia in emergency sclerotherapy of esophageal varices. Gastroenterology 1991; 101:1642-1648.
4. Rodriguez W, Levine J. Endteroccol endocarditis following flexible sigmoidoscopy. West J Med 1984; 140:951-953.

5. ASGE Antibiotic prophylaxis for gastrointestinal endosocpy. Gastrointest Endosc 1995; 42:630-635.
6. Dajani AS, Taubert KA, Wilson W et al. Prevention of bacterial endoscarditis. Recommendations by the American Heart Association. JAMA 1997; 277:1794-1801.
7. Meyer GW. Endocarditis prophylaxis for esophageal dilation: A confusing issue? Gastrointest Endosc 1998; 48:641-643. *Editorial.*
8. Malone J, Moore W, Campagna G et al. Bacteremic infectability of vascular grafts: The influence of pseudointimal integrity and duration of graft function. Surg 1975; 78:211-216.
9. Byl B, Deviere J, Struelens MJ et al. Antibiotic prophylaxis for infectious complications after therapeutic endoscopic retrograde cholangio-pancreatography: A randomized, double-blind, placebo-controlled study. Clin Infect Dis 1995; 20:1236-1240.
10. Niederau C, Pohlman U, Lubke H et al. Prophylactic antibiotic treatment in therapeutic or complicated diagnostic ERCP: Results of a randomized controlled clinical study. Gastrointest Endosc 1994; 40:533-537.
11. Sauter G, Grabein B, Huber G et al. Antibiotic prophylaxis of infectious complications with endoscopic retrograde cholangiopancreatography. A randomized controlled study. Endoscopy 1990; 22:164-167.
12. Van den Hazel SJ, Speelman P, Dankert J et al. Piperacillin to prevent cholangitis after endoscopic retrograde cholangio-pancreatography. A randomized, controlled trial. Ann Intern Med 1996; 124:442-447.
13. Finkelstein R, Yassin K, Suissa A et al. Failure of cefonicid prophylaxis for infectious complications related to endoscopic retrograde cholangiopancreatography. Clin Infect Dis 1996; 23:378-379.
14. Motte S, Deviere J, Dumonceau JM et al. Risk factors for septicemia following endoscopic biliary stenting. Gastroenterology 1991; 101:1374-1381.
15. Leung JWC, Ling TKW, Chan RCY et al. Antibiotics, biliary sepsis and bile duct stones. Gastrointest Endosc 1994; 40:716-721.
16. Jain N, Larson D, Schroeder K et al. Antibiotic prophylaxis for percutaneous endoscopic gastrostomy. Ann Int Med 1987; 107:824-828.

CHAPTER 4

Principles of Endoscopic Electrosurgery

Rosalind U. van Stolk

Introduction

Electrosurgery is the use of high-frequency alternating electric current to cut, coagulate or vaporize tissue. Electrosurgical techniques are used in endoscopy for removal of polyps, control of bleeding, enlarging the biliary sphincter (sphincterotomy) and tissue ablation.

- **Electricity—The Basics**
  Electric current flows through a wire like water flows through a pipe. A pump can exert force on the water to make it flow. The stronger the force, the faster the water flows. The force needed to make electrons flow is called voltage. Current is the measure of the rate of electron flow through a wire and is measured in amperes (amps). The amount of water that can flow through a pipe is affected by the diameter and the friction within the pipe. Similarly, electrons encounter resistance either due to the characteristics of the wire or the tissue through which the current is flowing. Electric power is measured in watts and is calculated from the force and the rate of flow. [Power (watts) = Force (volts) x rate (amps)] Electric current must complete a circuit in order to flow.
- **Alternating Current**
  Alternating current reverses direction continuously and is measured in cycles per second using units called hertz. Low frequency current such as that from a wall socket (60 hertz) can produce a shock. High frequency current in the range of 350 thousand to 1.5 million hertz has little effect on nerves and muscles, does not cause a shock. The primary effect if to heat tissue.

Current Characteristics and Tissue Effect

- **Current Density**
  The concentration of current passing through a given cross-sectional area of tissue. The higher the current density, the higher the heat and the greater the tissue effect Heat can be increased by either increasing the current or decreasing the cross-sectional area.
- **Waveform Settings**
  - Cutting mode—continuous sine wave with high current density. High temperature is generated which vaporizes cells, cuts through tissue and produces little coagulation. Voltages frequently are several hundred.
  - Coagulation mode—bursts of high-frequency waves with rapid fall-off. Lower current density and lower temperature causes desiccation of cells. Voltages typically are several thousand.
  - Blended current—Combined cutting and coagulation waveforms. Effect is cutting action with hemostasis.

Gastrointestinal Endoscopy, edited by Jacques Van Dam and Richard C. K. Wong.

Monopolar Electrosurgery

- **Principles**

  A monopolar circuit passes current from a small electrode, through the patient to a large dispersive electrode (pad) on the patient's skin. Because the return electrode is applied to such a large surface area, no heat is generated at this surface and all the electrosurgical effect is at the "active" electrode, which is passed down the endoscope.
- **Equipment**
  - Generator with cut, coag and blend setting
  - Dispersive electrode (commonly but erroneously called the "grounding" pad)—commercially packaged with a gel matrix
  - Active electrode ("hot biopsy" forceps, snares, sphincterotomes)
- **Techniques**
  - Hot biopsy
    - Commonly used to remove very small polyps
    - Grasp small polyp with forceps and tent tissue away from underlying wall creating a thin pseudo-stalk
    - Use blend or coag current in short bursts until pseudo-stalk is whitened
    - Complete polypectomy with quick tug avulsing the polyp off the mucosa
    - Tissue in forceps jaws is sent to pathology lab for evaluation
  - Snare
    - Used to remove larger polyps
    - Pass snare in sheath down accessory channel and a few mm into lumen
    - Deploy snare out of sheath and pass snare loop around neck of polyp
    - Snug snare closed taking care not to cut through stalk with the snare wire
    - Use blended or coag current in short bursts while assistant continues to slowly close the snare
    - Retrieve specimen for pathologic evaluation
  - Sphincterotomy
    - Used to open the biliary sphincter to allow access to the biliary and pancreatic duct for fluoroscopy as well as stone retrieval and cholangioscopy
    - Pass closed spincterotome down accessory channel of duodenoscope and position into the ampulla
    - Tighten the wire until it bows sightly and the midpoint of the wire is across the ampullary opening
    - Use short bursts of current to cut and cauterize the sphincter
- **Complications**
  - Dispersive electrode burns
    - Area of skin contact too small
    - Gel has dried
    - Air gaps between pad and skin
  - Alternate site burns
    - Can occur whenever an alternate current path is present
    - Tip of forceps or snare sheath not fully out of endoscope tip
    - Active electrode in contact with normal mucosa
  - Deep tissue injury
    - Current applied to too large an area of tissue
    - Current applied too long

- Inadequate creation of pseudo-stalk (hot-biopsy)
- Use of monopolar forceps as a contact probe with pressure applied
- Stray currents
  - Electric shock
  - Electrical interference
  - Patient or operator burns

- **Precautions**
  - Implanted Cardiac Devices
    - Deactivate implantable cardioverter defibrillator (ICD) during electrosurgical use
    - Apply current in short repetitive bursts in pacemaker patients
    - Use bipolar devices if possible in pacemaker patients
    - Monitor cardiac rhythm carefully throughout procedure
    - Position dispersive electrode distant from the cardiac device or leads

Bipolar Electrosurgery

- **Principles**
  - Two active electrodes are separated by an insulator
  - Both electrodes come in contact with tissue
  - No dispersive electrode is necessary
  - No current passes through patient
- **Equipment**
  - Bipolar contact probes are available from many vendors and are commonly used to control bleeding
  - Giant contact probes have been used to open obstructing esophageal and rectal cancers
  - Bipolar snares and sphincterotomes are available but are not commonly used
- **Technique**
  - Advance tip of probe with electrodes fully out of endoscope tip
  - Apply probe tip to edge of bleeding area with moderate pressure and activate current
  - Treated tissue should turn white-gray
  - Large bleeding areas should be treated concentrically toward the center
- **Limitations**
  - Inadequate tissue effect
    - Operating in a wet field
    - Tissue coagulum adhering to probe tip
  - No bipolar hot biopsy application
    - Bipolar forceps destroy all tissue in forceps cups leaving no ability to assess tissue histologically

Non-Contact Electrosurgery (Argon Plasma Coagulation)

- **Principles**
  - Current applied to tissue through ionized argon gas
  - Electric circuit is completed from the probe tip, through the gas to the closest tissue with the lowest impedance
  - Coagulation or desiccation effect on tissue
  - Direction of current is not dependent upon the direction of gas flow or the positioning of the probe tip

- **Equipment**
  - Argon gas source
  - High frequency current generator
  - Dispersive electrode pad
  - Argon gas probe
  - Finger and foot switches for simultaneous activation of the gas source and the current generator
- **Indications**
  - Treatment of diffuse vascular lesions such as radiation proctitis and antral vascular ectasia
  - Ablation of remaining small amounts of polyp tissue after polypectomy
- **Technique**
  - Placement of probe near but not touching target tissue
  - Treatment with a "painting" or stroke technique
- **Limitations**
  - High voltages necessary to ionize gas therefore adequate insulation of leads and probes is essential
  - All safety precautions necessary for monopolar electrosurgery must be followed
  - Significant and often insurmountable electrical interference affecting view of the video screen is common

Selected References

1. Tucker R. The physics of electrosurgery. Continuing education for the Family Physician. 1985; 20(8):574-589.
2. Williams CB. Diathermy-biopsy-A technique for the endoscopic management of small polyps. Endoscopy 1973:215-218.
3. Goodall RJ. Bleeding after endoscopic sphincterotomy. Ann Roy Coll Surg Engl 1985; 67:87-88.
4. Wadas D, Sanowski R. Complications of hot biopsy forceps technique. Gastrointest Endosc 1988; 34:32-37.
5. Technology Assessment Status Evaluation: Electrocautery use in patients with implanted cardiac devices. Gastrointest Endosc 1994; 40:794-795.
6. Verhoeven AGM, Bartelsman JFWM, Huibregtse K et al. A new multipolar coagulation electrode for endoscopic hemostasis In: Proceedings of the 13th International Congress on Stomach Diseases. Amsterdam-Oxford, Princeton: Excerpta Medica 1981:216-221.
7. Papp J. Endoscopic treatment of gastrointestinal bleeding: Electrocoagulation. In: Sivak, ed. Gastroenterologic Endoscopy. W.B. Saunders, 1987.
8. Farin G, Grund KE. Technology of argon plasma coagulation with particular regard to endoscopic applications. Endoscopic Surgeries and Allied Technologies. 1994; 2(1):71-77.

CHAPTER 5

The Benign Esophagus

James M. Gordon

Anatomy

- The esophagus serves as a conduit for food as it traverses from the mouth to the stomach. The distal two-thirds of the esophagus consists of smooth muscle, and the proximal one-third is primarily skeletal muscle. The normal diameter is 15-20 mm, and the length of the esophagus is 20-25 cm, spanning the cricopharyngeus muscle to the diaphragmatic hiatus. An EGD begins when the endoscope is passed through the incisor teeth, over the tongue, to the level of the cricopharyngeus muscle, which helps form the upper esophageal sphincter (UES). At the UES there is a slit like opening, that with swallowing, opens and allows passage of the endoscope. The normal mucosal color of the esophageal squamous epithelium is pale pink. Often secondary peristaltic contractions can be observed, as the endoscope passes distally. At 25-30 cm from the incisors, the aorta can be seen compressing the esophagus and just distal to this an impression can be seen from the left main stem bronchus. In the distal esophagus, there is a clear demarcation between the two types of mucosae, that often appears as a line, called the squamocolumnar junction or Z line (ora serrata). The Z line is normally located at the level of the lower esophageal sphincter (LES), is within the distal 2 cm of the esophagus, and marks the transition from squamous to columnar mucosa. As the endoscope is passed into the stomach, the LES opens spontaneously, allowing unimpeded passage into the stomach. Once the endoscope is passed into the stomach, the esophageal exam is completed with retroflexion to examine the cardia of the stomach.

Esophagitis

- Gastroesophageal reflux disease (GERD). GERD is the most common disease of the esophagus, accounting for approximately 75% of esophageal disorders. The lifetime prevalence of GERD is 25-35%, and a recent survey indicated 44% of Americans have heartburn once a month, and 7% take antacids daily. The majority of reflux patients, however, never present to their physician and either make lifestyle modifications or use over-the-counter medications.
  - Typical and atypical manifestations. The typical manifestations of GERD include a burning substernal sensation (heartburn) that radiates to the jaw, belching, nausea, water brash or hypersalivation. These symptoms are typical if they are relieved with standard medications and antacids. Recently, there has been increased interest in the atypical manifestations of GERD. These include chest pain not of cardiac origin, chronic cough, asthma, hoarseness, dental erosions, pulmonary fibrosis, bronchiectasis and dyspepsia. Up to 75% of noncardiac chest pain is caused by underlying gastroesophageal reflux (GER), but only 10% of these patients actually have esophagitis. 33-80%

Gastrointestinal Endoscopy, edited by Jacques Van Dam and Richard C. K. Wong.

Table 5.1. Pathophysiology of GERD

Antireflux Barrier Abnormalities	
	Inappropriate transient lower esophageal sphincter relaxations (TLESRs)
	Hiatal hernia
	Hypotensive lower esophageal sphincter (less than 10%)
	Overwhelmed LES (Increased intra-abdominal pressure)
Esophageal Motility Abnormalities	
	Impaired peristalsis
	Decreased saliva
Gastric Factors	
	Increased acidity (Barrett's esophagus only)
	Increased food volume
	Dietary factors
Other Factors	
	Cigarette smoking
	Hormones
	Medications

of cases of asthma, chronic cough are exacerbated by GERD. Studies of dental erosions and GERD show that 83% of patients with dental erosions have GERD as documented by 24 hour pH testing, and 55% of patients with GERD have dental erosions. 33% of patients with an atypical presentation of GERD may have no heartburn symptoms at all.

- **Pathophysiology**. The pathophysiology of GERD is summarized in Table 5.1. Transient lower esophageal sphincter relaxation (TLESRs) is the most common mechanism through which gastric contents are refluxed into the esophagus. 94% of all patients with GERD have inappropriate TLESRs and less than 15% have a hypotensive LES, with most patients having normal LES pressures (15-37 mm Hg). A hiatal hernia predisposes the patient to reflux, through disruption of the anti-reflux barrier (LES plus diaphragm), with the LES and a gastric pouch in the chest, and the diaphragm in its normal anatomic position, providing a free conduit for food and acid to reflux into the esophagus.
- **Evaluation.**
  - Barium swallow. The barium esophagram is the test of choice for dysphagia. It can reliably detect strictures, webs, and diverticula. It can assess the emptying capacity of the esophagus in order to make sure that achalasia or scleroderma are not present. As a test for GERD, however, it has a sensitivity of only 40%. Additionally, a normal esophagram does not preclude the need for endoscopy.
  - **Upper endoscopy**. An EGD allows the examiner to assess mucosal status as well as to biopsy the esophagus in order to diagnose GERD. The reason to perform an EGD is to identify those patients with Barrett's esophagus, as these patients need regular endoscopic screening. Esophagitis, either macroscopic or microscopic, confirms the diagnosis of GERD. Therefore in patients in whom an endoscopic procedure is being performed for GERD, if the mucosa appears normal, the endoscopist should biopsy the distal esophagus to look for microscopic tissue damage. If macroscopic

Table 5.2. Representative endoscopic grading system for esophagitis

Hetzel-Dent Classification	
Grade 0	No mucosal abnormalities
Grade 1	No macroscopic lesions, but erythema, hyperemia, or mucosal friability
Grade 2	Superficial erosions involving < 10% of the mucosal surface of the last 5 cm of esophageal squamous mucosa
Grade 3	Superficial erosions or ulcerations involving 10% to 50% of the mucosal surface of the last 5 cm of esophageal squamous mucosa
Grade 4	Deep peptic ulceration anywhere in the esophagus or confluent erosions of > 50% of the mucosal surface of the last 5 cm of esophageal squamous mucosa.
Savary-Miller Classification	
Grade I	Linear, nonconfluent erosions
Grade II	Longitudinal, confluent, noncircumferential erosions
Grade III	Longitudinal, confluent, circumferential erosions that bleed easily
Grade IVa	One or more esophageal ulcerations in the mucosal transition zone which can be accompanied by stricture or metaplasia.
Grade IVb	With the presence of stricture but without the presence of erosions or ulcerations

esophagitis is seen on EGD, a repeat endoscopy should be performed after the esophagitis is healed because Barrett's esophagus or dysplasia is difficult to diagnose with esophagitis.

- **Incidence of esophagitis.** The incidence of esophagitis in patients with reflux demonstrated on pH study is 54%. However, the degree of the symptoms and the extent of esophagitis do not always correlate. Esophagitis can range from mild erythema in the distal esophagus to exudates. Endoscopic grading of esophagitis is important due to therapeutic and prognostic implications, with lower healing rates and more aggressive therapy necessary in patients with severe esophagitis. Several grading systems have been proposed. The two most popular grading systems in use are the Hetzel-Dent classification and the Savary-Miller classification system both of which are outlined in Table 5.2.
- **Barrett's esophagus.** Barrett's esophagus is a condition in which the squamous mucosa lining the distal esophagus is changed to specialized columnar epithelium with intestinal metaplasia. Endoscopically, the squamocolumnar junction is seen more than 3 cm above the lower esophageal sphincter. Often this appears as salmon-pink mucosa that can be finger projections into normal mucosa, islands of pink mucosa or complete replacement of the squamous mucosa with Barrett's epithelium. The diagnosis is confirmed on microscopic examination, with intestinal metaplasia and goblet cells seen on biopsy. Barrett's esophagus is seen in 10-15% of all patients undergoing an upper endoscopy. Its causes are multifactorial, but it is clearly linked to GERD. The significance of Barrett's esophagus is that patients are predisposed to develop adenocarcinoma of the esophagus, which is the fastest rising incidence of any cancer in the United States at present. The incidence of adenocarcinoma has been estimated to be 1 cancer per 100 years of

follow-up. Given the increased cancer risk, patients with Barrett's esophagus need to have screening upper endoscopies performed at regular intervals, in order to screen for dysplasia. Patients who are newly diagnosed should have a screening EGD 2 years after initial diagnosis and every 2 years thereafter. If dysplasia is found, the finding needs to be confirmed by an expert pathologist. For patients found to have low grade dysplasia, aggressive antireflux therapy with PPIs should be started, with repeat EGD in 8 to 12 weeks to obtain multiple biopsy specimens. If the repeat exam shows no evidence of dysplasia, an EGD should be performed every 6 months until two consecutive negative exams. If the repeat study shows persistent low grade dysplasia, aggressive antireflux therapy and repeat EGD at 6 month intervals are indicated. For high grade dysplasia, surgery with esophagectomy remains the mainstay recommendation, although initial data with new approaches such as photodynamic therapy (PDT) have been promising. In patients found to have Barrett's esophagus with no dysplasia, aggressive management of GERD with proton pump inhibitors is indicated, although regression of Barrett's epithelium has never been clearly shown.

- Strictures, webs and rings. Narrowing of the esophagus can occur for various reasons. The most common reason for luminal narrowing is a gastroesophageal reflux. Table 5.3 lists the common causes of esophageal stenosis. **Esophageal webs** and **rings** are thin, fragile mucosal bands that interrupt the lumen of the esophagus. A ring is a mucosal structure that marks the gastroesophageal junction and is covered on the distal side by columnar epithelium and on the esophageal side by squamous epithelium. Endoscopically, a thin mucosal band can be seen. The most common ring is a lower esophageal ring or Schatzki's ring that almost always occurs in conjunction with a hiatal hernia. Symptomatic patients will experience intermittent dysphagia. Etiology of Schatzki's ring is controversial, although there is evidence that some of these may be caused by chronic gastroesophageal reflux.

An **esophageal web** is a mucosal band that is covered on both sides by squamous epithelium, and can occur anywhere above the squamocolumnar junction. The etiology of webs is unclear, although there are associations with other disorders. In the Plummer-Vinson syndrome a cervical web occurs in association with iron deficiency anemia. Cervical webs may also be seen in association with dermatologic disorders such a epidermolysis bullosa and benign mucous membrane pemphigoid. Mid-esophageal webs have been described not only with the above dermatologic conditions but also in chronic graft-versus-host disease and with psoriasis. Treatment of webs and rings consists of mechanical disruption of the ring. This can occur with passage of the endoscope, or with bougiennage if necessary.

Peptic strictures can occur in patients with chronic GERD. These patients often present with dysphagia, but can also be asymptomatic depending on the degree of luminal narrowing. Endoscopically, strictures can range from thin fibrous bands that slightly narrow the lumen to long thin areas that cannot be traversed by an endoscope. Strictures found on a radiographic study must be evaluated with an upper endoscopy and biopsy. Treatment includes dilation, often requiring more than one session, depending on the lumen diameter and response to therapy. Peptic strictures also need to be

5

Table 5.3. Etiology of esophageal narrowing

Intrinsic Stenosis	
	GERD
	Tumors
	Rings and webs
	Epidermolysis bullosa
	Benign mucous membrane pemphigoid
	Chronic graft-versus-host disease
	Infections
	Caustic ingestion
Extrinsic Compression	
	Tumors of lung or mediastinum
	Mediastinal infections
	Vascular anomalies
Iatrogenic Causes	
	Radiation
	Sclerotherapy
	Nasogastric tubes
	Pill-induced injury

treated with proton pump inhibitors, as there is data to suggest that the use may prevent recurrence.

The **most common symptom** that the patient with esophageal luminal narrowing experiences is dysphagia. All patients in whom the diameter of esophagus is less than 13 mm will experience dysphagia. Initial evaluation should consist of an esophagram with a 13 mm barium pill, because this is sensitive for demonstrating subtle esophageal narrowing or specific regions causing a holdup of bolus transit. Radiography, however does not reliably distinguish between benign and malignant strictures. Nor does it treat webs, rings or peptic strictures. Therefore endoscopy should be done, along with biopsy and brushings of the stricture. Additionally if there is further question regarding the nature of the stricture, endoscopic ultrasound can be done to detect tissue invasion and depth.

The **treatment of benign strictures**, such as those caused by gastroesophageal reflux, consists of aggressive use of proton pump inhibitors and mechanical dilation of the structured area. There are three basic types of devices that are in use today. **Mercury filled rubber bougies** (Maloney dilators) are useful for dilating uncomplicated strictures, ones that are not very tight, nor irregular in shape and contour. **Fixed-size dilators** that are passed over a guide wire (Savary-Gilliard dilators) under fluoroscopic guidance can be used for complicated strictures that are so narrow that the endoscope is unable to be passed through it, as well as those that appear irregular. The technique involves passing the endoscope to the stricture and passing a guidewire under fluoroscopic guidance through the narrowed segment into the stomach, then leaving the wire in place. The endoscope is withdrawn and the Savary dilator is passed over the guidewire beyond the strictured area into the stomach.

Through the scope (TTS) dilators are useful for those strictures that require endoscopic guidance. The dilator is passed into the middle of the stricture through the biopsy channel of the endoscope. The balloon is then inflated to the appropriate pressure for 60 sec while the endoscopist insures that the dilator remains in place. There is no clear consensus as to the optimal size which a stricture should be dilated, and the patients symptomatic response may be the best guide. About 40% of patients experience complete relief of dysphagia with one dilation, and others need repeated sessions, often occurring for months, or longer.

- **Hiatus hernia.** There are two major types of hiatal hernias: a sliding hernia and a paraesophageal. The **sliding hiatal hernia** is the most common type. With a sliding hiatal hernia, both the esophagogastric junction and part of stomach migrate through the diaphragmatic hiatus and reside in the posterior mediastinum. Sliding hiatal hernias are often associated with GERD and may contribute to the development of reflux esophagitis through disruption of the normal antireflux barrier. Although patients with hiatal hernias may have GERD, the hernias themselves are rarely symptomatic. A **paraesophageal hernia** occurs when the esophagogastric junction is in its normal position and the fundus of the stomach ruptures the diaphragmatic hiatus into the chest. Therefore a portion of the stomach may be positioned alongside the esophagus. The risk of a paraesophageal hernia is that a gastric volvulus (twisting) can occur, and should vascular compromise occur, gastric infarction and perforation are possible. Therefore when discovered, it is recommended that all paraesophageal hernias undergo elective surgical correction.
- Esophageal manometry. The role of **esophageal manometry** in the diagnosis of GERD is very limited. Manometry is useful to assess the tone and function of the lower esophageal sphincter, as well as to assess for the adequacy of peristalsis. It should be done before pH testing, as it is the only accurate method to detect the location of the LES. Manometry is also an integral part of the preoperative evaluation prior to anti-reflux surgery, to rule-out other diagnoses (e.g., achalasia and scleroderma), as well as to assess the adequacy of esophageal body peristaltic contractions.
- pH testing and its role in GERD management. Esophageal **pH testing** is the gold standard for diagnosing reflux disease. The test is done by placing a small catheter via the nose into the esophagus in order to detect the percentage of time the pH is less than 4.0 in the distal esophagus. The distal esophagus is defined as the spot that is 5 cm above the manometrically identified LES. The advantage of pHmetry is that not only is the amount of reflux quantified, but the patients symptoms can be correlated with episodes of reflux. It needs to be emphasized that an accurate test should be done under normal living conditions and that all efforts need to be made to do the activities that induce reflux symptoms. Normal values for distal and proximal reflux are shown in Table 5.4. The test can be done on medications if the diagnosis of reflux is known and established. A 24 hour test should be done off of medications if the diagnosis of pathologic reflux is in doubt.
- **Treatment.** In a patient who has short duration symptoms, which occur only several times per week, an empiric trial of therapy is warranted. Empiric therapy should always consists of lifestyle modifications (see Table 5.5) as well as a trial of appropriate medication. If empiric therapy is used, a careful history is necessary to determine if alarm symptoms are present: weight

Table 5.4. Representative normal values for 24 hour pH tests

Sensor	Measure	Value
Distal Sensor	Total Time pH < 4:	< 5.5%
	Supine Time pH < 4:	< 3.0%
	Upright Time pH < 4:	< 8.2%
Proximal Sensor	Total Time pH < 4:	< 1.1%
	Supine Time pH < 4:	< 0.6%
	Upright Time pH < 4:	< 1.7%

5

loss, anemia, occult blood in the stool, dysphagia, or odynophagia. If the patient's symptoms are longer than one year, or disabling, then the patient should undergo endoscopy. Caucasian males are at most risk for Barrett's esophagus and should have an immediate diagnostic evaluation. If over the counter medications are not effective, then using a prescription strength histamine type 2 receptor antagonist (cimetidine, ranitidine, famotidine, nizatidine) or a prokinetic agent (cisapride) is indicated. Doses of the H2RAs need to be BID, with one report indicating that ranitidine needs to be used at 150 mg QID to be effective. 50-60% of patients will get relief of heartburn symptoms with therapy with H2RAs or cisapride. If symptoms are particularly severe or there is no response to initial therapy, then a proton pump inhibitor (omeprazole, lansoprazole, rabeprazole) can be used. If a PPI is used an endoscopy should be performed in order to look for Barrett's esophagus. PPIs provide relief in 75-80% in daily dosing and 20-25% in twice daily dosing. If esophagitis is seen on endoscopy, especially grades 2-4, a BID dose of PPI should be considered because longer acid suppression correlates with improved esophageal healing. Even BID dosing, however, does not completely suppress acid production. On omeprazole 20 mg BID, up to 70% of volunteers were found to resume acid production after bedtime. Therefore maximal acid suppression is achieved by giving a PPI twice daily, before meals, and an H2RA before bedtime. Acute side effects of the PPIs are minimal (headache, diarrhea and abdominal pain), but the long-term safety has been a concern in the past . These drugs have been associated with hypergastrinemia and proliferation of the enterochromaffin-like cells. Prolonged therapy has been associated with the development of carcinoid tumors in rats, but not in mice or dogs. However, long term studies in humans as well as long term clinical use show that the risk of developing a carcinoid tumor is the same as in the general population. Thus these drugs appear to be safe for long-term use and are currently approved by the FDA for prolonged use.

- **Chronic management.** GERD is a chronic, relapsing condition. Those with an endoscopically normal esophagus may need only as needed or over the counter medications. 80% of patients with **esophagitis relapse** rate at one year of discontinuing therapy. The major predictors of relapse appear to be the initial grade of esophagitis, resistance to healing with an H2RA, and a low LES pressure. Therefore, many patients will need maintenance therapy. The doses used for maintenance are similar to that of acute disease. Ranitidine

Table 5.5. Lifestyle modifications in the treatment of gastroesophageal reflux disease

- Elevate the head of the bed
- Stop Smoking
- Decrease alcohol consumption
- Decrease fat intake
- Decrease meal size
- Avoid bedtime snacks
- Weight reduction if overweight
- Avoid certain foods: Caffeinated products
Chocolate
Peppermint
Citrus products
- Avoid certain drugs: Theophylline
Beta agonists
Calcium channel blockers
Narcotics

150 mg BID is approved for maintenance GERD therapy, and several studies suggest cisapride 20 mg BID may be effective for mild esophagitis. About 50% of cases long-term PPIs are needed. One year studies in patients with endoscopic esophagitis show 80% of patients remain in remission, as compared to 20 to 40% with ranitidine and less than 20% remission with placebo. One study showed that at 5 years 100% of patients were in remission on omeprazole, but a small percentage of these patients needed to be on either 40 mg or 60 mg per day for maintenance.

- **Anti-reflux surgery**. This procedure aims to reduce a hiatal hernia as well as wraps the distal end of the esophagus with the fundus of the stomach (Nissen fundoplication). The wrap has been shown to increase the basal LES pressures, decreases the frequency of transient LES relaxation and inhibit complete LES relaxation. The fundoplication wrap may be 360° (Nissen) or 180°-270° (Toupet). 90% of the patients are in remission at one year with the laparoscopic approach. Considerable controversy exists over the long-term effectiveness of surgery for GERD. In older literature, two studies that compare open fundoplication with either H2RAs, metoclopramide or antacids clearly indicate that surgery is superior in providing long-term relief. There are no published trials comparing laparoscopic anti-reflux surgery with omeprazole. One preliminary report from Scandinavia indicates an identical response at 3 years for laparoscopic surgery and omeprazole 20 mg daily. Critical for successful anti-reflux surgery is selecting the right patient as well as the best surgeon. The ideal surgical candidates are those that respond to medical therapy. If a patient has extraesophageal symptoms prior to surgery, that are not responsive to aggressive medical management, then these symptoms will likely persist postoperatively. Patients considered for surgery should also undergo esophageal manometry to make sure the peristaltic amplitude is normal (> 30 mm Hg), as well as a 24 -hour pH test, endoscopy and gastric emptying study in select patients.

- **GERD treatment failures.** Important issues in GERD are defining a medical failure and when to consider an antireflux procedure. The most common reason that patients fail medical therapy is that they are under treated. Patients should not be referred for surgery due to lack of response to medications, as all patients should have a complete response to medications, even if large doses are needed. Table 5.7 lists the most common reasons patients are termed GERD treatment failures. In patients who do not respond to therapy clinical reappraisal is important to insure the diagnosis of GERD is correct. In patients who develop esophagitis on medications, especially those with discrete ulcers, a careful look at the medication list can help to rule out pill-induced injury. Drugs such as alendronate, doxycycline, nonsteroidal anti-inflammatory drugs and potassium chloride have all been associated with esophageal mucosal injury. The most common sites of **pill esophagitis** are the distal esophagus and the area where the aortic arch passes over the esophagus. Treatment involves removing the offending agent.

- **Management of atypical manifestations of GERD.** The atypical symptoms of GERD can often be difficult to recognize and to treat. For the extraesophageal manifestations such as cough, asthma and hoarseness, a trial with at least twice daily high dose proton pump inhibitors (i.e., lansoprazole 30 mg BID) for at least 3 to 6 months should be considered prior to formal testing. To maximize therapy, consideration should be given to adding a bedtime H2RA. Aggressive medical therapy can be used as a diagnostic test: if the patients improve, the diagnosis of GERD-related symptoms is proven; if the patients fail therapy, then a 24-hour pH study should be done on medications to assess for efficacy of therapy as well as to determine if the symptoms correlate with episodes of reflux.

  In patients with noncardiac chest pain, it is important to make sure that there are no cardiac abnormalities that can be causing the symptoms. Once that is proven either a short diagnostic trial with a high dose twice daily proton pump inhibitor can be done or a 24-hour pH test can be done to determine if the symptoms and reflux correlate.

- Infectious causes of esophagitis
  - Fungal infections. Endoscopy the test of choice for diagnosing esophageal **fungal infections.** It helps in distinguishing Candida infections form herpetic esophagitis, which can often mimic each other. The most common fungal infection is *Candida albicans*. The endoscopic findings in Candida esophagitis are nonspecific, although highly characteristic. Most often, there are patches of a cream colored pseudomembrane, but this can range form highly friable, erythematous mucosa to pseudomembranes covering the entire mucosa. Any segment of the esophagus can be involved, but it is most common in the lower two-thirds of the esophagus. Diagnosis can be suspected based on the endoscopic appearance, but biopsy and brushings are the most accurate diagnostic method. The endoscopist can confirm the diagnosis by doing a 10% KOH prep and a NaCl wet prep to look for fungi under the microscope after the endoscopic procedure is complete. Treatment is with an antifungal agent such as fluconazole 100 mg daily or ketoconazole 200 mg daily for 7 days.

Table 5.6. GERD treatment failures

- Under treatment (not enough acid suppression)
- Pill esophagitis
- Incorrect diagnosis (achalasia, visceral sensory afferent dysfunction)
- Hypersecretion of acid (Zollinger-Ellison syndrome)
- Delayed gastric emptying
- Bile reflux (?)

- Herpes simplex esophagitis. **Herpetic esophagitis** is most common in immunocompromised patients but occasionally can be seen in normal patients. Empiric treatment of herpes esophagitis can be undertaken in a patient with the typical symptoms of chest pain, dysphagia and oropharyngeal herpetic lesions. The endoscopic appearance of HSV esophagitis reflects the stages through which the virus progresses with initial stage I, small vesicular lesions with an erythematous base. Stage II, the vesicles progress to "volcano" ulcers which are shallow with heaped up edges. In stage III, the adjacent ulcers coalesce into larger ulcers with areas of exudate, hemorrhagic areas and friable epithelium. Unless the vesicular lesions are present, biopsy is necessary to distinguish these ulcers from other causes of esophageal ulcers. Biopsies for HSV are best obtained from the edges of the ulcer. Characteristic histologic findings of HSV include ground-glass nuclei, **mulitnucleate giant** cells and Cowdry type A intranuclear inclusions. Treatment for HSV esophagitis is with acyclovir 200 mg 5 times daily and in refractory patients ganciclovir (5 mg/kg IV BID) or foscarnet (90 mg/kg IV BID) can be used.
- Cytomegalovirus esophagitis. **Cytomegalovrus esophagitis** is seen only in immunocompromised patients. It is distinguished from HSV esophagitis by having no vesicular stage. The early ulcers of CMV are serpiginous and eventually form giant ulcers in the distal esophagus. Biopsies are taken from within the ulcer base (as compared to HSV), as it is very difficult to make the diagnosis based on the endoscopic appearance alone. The pathology reveals large cells in the submucosa with intranuclear inclusions. Treatment is with either foscarnet (90 mg/kg IV BID) or ganciclovir (5 mg/kg IV BID).
- Special considerations for the acquired immunodeficiency syndrome. Esophageal involvement with AIDS is frequent. The most common cause of esophagitis in **AIDS** is Candida. Prospective endoscopic studies evaluating esophageal complaints in AIDS have shown that 50-79% of subjects will have esophageal candidiasis alone or in association with other causes. The most common causes of ulcerative esophagitis are CMV, followed by idiopathic esophageal ulcers (IEU). HSV esophagitis in AIDS is actually relatively uncommon, as compared to hosts with immunosuppression for other reasons. Multiple processes can occur in 10% of patients. Pill esophagitis (due to AZT or ddC) and GERD can also be seen.

  The most common symptoms that patients experience are odynophagia or dysphagia. Heartburn and regurgitation may also be seen and may reflect an opportunistic infection, but the majority of cases with these symptoms reflect underlying GERD. When evaluating patients with esophageal complaints, the most important part of the physical exam is examination of the oral cavity. Two-thirds of patients with esophageal candidiasis have oropharyngeal candidiasis. Patients with CMV or IEU esophagitis, rarely have oral mucosal lesions, but HSV typically does.

Given that esophageal candidiasis is the most likely cause of esophageal disease in AIDS, it is recommended that an empiric trial be given for one week and that further diagnostic testing depend on clinical response. Most patients will get marked symptomatic improvement by day 3, and symptoms should be completely resolved by day 7. If the symptoms do not improve, then endoscopy is indicated. This approach was recently evaluated and shown to be safe and cost effective in a prospective trial (Wilcox et al. Gastroenterol 1996; 110: 1803-1809). If the patient has a subsequent relapse, however, it is advisable to endoscope the patient prior to starting empiric therapy. In patients with ulcers, biopsies are essential to distinguish between CMV and IEU. At endoscopy biopsies should be obtained of any abnormalities. At least six biopsies should be obtained from each ulcer. If patients with adequate numbers of biopsies are obtained and no specific cause is found, IEU should be considered and the patient started on therapy. Treatment for IEU is with either prednisone 40 mg daily tapered over 4 weeks or thalidomide 90 mg/kg BID.

- Esophageal motility disorders. Esophageal motility disorders mainly manifest through the symptoms of chest pain or dysphagia. GERD is the underlying cause of most esophageal motility problems and needs to be ruled out. However, disorders such as diffuse esophageal spasm (DES), achalasia, and scleroderma represent motility disorders that probably are not alone due to GERD. A classification of esophageal motility disorders is presented in Table 5.7.
  - **Achalasia.** Achalasia is the pure esophageal motility disorder in which the LES becomes aganglionic and is thus incapable of relaxing with swallows. The incidence has been estimated at 1/100,000 population per year, affecting both sexes equally. Presenting symptoms include regurgitation, dysphagia and chest pain. Usually patients have had these symptoms for a long time and have adapted well to them. The average time to diagnosis for achalasia is 7 years, and some patients may go as long as 20 years before the correct diagnosis is made. The diagnosis is made by a combination of a barium esophagram and esophageal manometry. On esophagram, the characteristic findings are a dilated esophageal body with an air-fluid level and tapering of the LES region referred to as a "birds beak". Patients with long-standing disease, a "sigmoid" esophagus is seen that denotes the esophagus is no longer functional. In some cases, an air fluid level can be seen on plain chest radiograph as well. Esophageal manometry shows no evidence of peristalsis, a hypertensive LES pressure that does not completely relax, and an elevated esophageal body baseline pressure. The role of endoscopy for diagnosis of achalasia is to examine the esophagus as well as to obtain biopsies of the gastric cardia to rule out adenocarcinoma of the cardia which is the most common cause of pseudoachalasia. Endoscopic clues to achalasia include a large amount of food and fluid retained in the esophagus, as well as a lack of peristalsis, a dilated esophageal body and a distal esophageal diverticulum. In one prospective study, achalasia was accurately diagnosed by endoscopy in only one-third of cases (Gut 1992; 33: 1011-1015).

Table 5.7. Classification of primary esophageal motility disorders

Absent peristalsis	Achalasia
Incoordinated motility	Diffuse esophageal spasm* Hypercontracting Esophagus "Nutcracker esophagus Hypertensive LES *
Hypocontracting esophagus	Ineffective esophageal motility (formerly referred to as nonspecific motor disorder)* Hypotensive LES*

*May be secondary to chronic GERD

-Treatment. No therapy for achalasia reverses the underlying neuropathology or associated impaired LES relaxation and aperistalsis. Instead, the target of therapy is to reduce LES pressure. **Medical therapy of achalasia** consists of the use of using smooth muscle relaxants, either calcium channel blockers or nitrates. Unfortunately medical therapy is ineffectual, providing relief to about 30% of patients. Yet it is a reasonable option to try prior to invasive management if the patient is elderly with concomitant medical problems. The standard therapy for achalasia consists of either **pneumatic dilation** or Heller myotomy. Both have about 85-90% initial relief and 80% long-term remission. A pneumatic dilation involves placing a large diameter balloon across the LES and forcefully ripping the muscle. An analysis of pneumatic dilation found that a post dilation LES pressure of less than 10 mm Hg was the best predictor of prolonged remission, whereas patients with a residual pressure of greater than 20 mm Hg had little benefit. A graded dilation should be done, using initially a 30 mm dilator, followed by a 35 and then a 40 mm dilator. Repeated dilation should be done depending on the patients' symptomatic response, although another method may be to look not only at the symptoms, but also at the esophageal emptying time. The main risk with pneumatic dilation is perforation, which can occur in 1-15% of cases, and therefore a gastrograffin esophagram must be done on all patients following dilation, prior to discharge. Any substantial perforation requires surgical repair. Perforations recognized within 6 to 8 hours have outcomes comparable to elective Heller myotomy. Success of therapy of achalasia is reported as relief of symptoms. With pneumatic dilation, 50-90% report good to excellent results. 30% of patients will require more than one dilation. The advantage to pneumatic dilation is that it is an outpatient procedure, recovery is rapid and discomfort is minimal.

The **surgical approach** to achalasia is to disrupt the LES enough to eliminate symptoms without causing excessive reflux. The current approaches are to perform the procedure either via a thoracotomy or laparotomy and to make an anterior myotomy. More recent trends are to perform this procedure laparoscopically, dividing the phrenoesophageal ligament and partially mobilizing the esophagus. Surgeons also debate whether to perform an anti-reflux procedure with the myotomy. The most popular approach at present is to perform a partial anterior fundoplication (Dohr procedure). Myotomy produces good to excellent results in 60-90% of patients and the operative mortality is low. When done through the minimally invasive approach, the hospital stay is three days, and the patient resumes regular activities in 10 to 14 days. Another therapeutic approach to achalasia is to inject intrasphincteric **Botulinum toxin**. This is done endoscopically, although some authors have argued that endoscopic ultrasound to localize the LES is optimal. 20 units of botulinum toxin is injected into each quadrant of the LES, with a sclerotherapy needle. In a trial comparing Botox to placebo, botox provided significantly better symptomatic, manometric and radiographic response, but 7 of the 8 patients relapsed in a mean of 7 months (Gastroenterology 1996; 111: 1418-1424). Comparing Botox to pneumatic dilation, at one year there was a sustained response in 32% ofpatients with botox, as compared to 70% with pneumatic dilation. Given that the response to botox is short-lived, botox should be reserved for those patients that present a high surgical risk, and can often be used as a bridge until the patients condition can be optimized to undergo pneumatic dilation. If patients relapse, they can be reinjected, with somewhat less response. The patients who respond best to Botox are older then age 50 and those with vigorous achalasia.

5

- **Scleroderma**. Esophageal involvement occurs in 70-80% of patients with scleroderma, with over 90% having associated Raynaud's phenomena. It is seen in both progressive systemic sclerosis and CREST (calcinosis, Raynaud's phenomena, esophageal involvement, sclerodactyly and telangiectasias). The pathophysiology involves an abnormality in muscle excitation and responsiveness due to muscle atrophy and decreased cholinergic innervation. The classic manometric features of advanced scleroderma include: low LES pressure, peristaltic dysfunction of the smooth muscle portion of the esophagus characterized by low amplitude contractions or aperistalsis, and preserved function of the striated muscle portion of the esophagus and oropharynx. As a result of these manometric abnormalities, patients may have dysphagia and severe GERD. Surprisingly, dysphagia for solids and liquids is reported by less than half of patients with scleroderma. More severe dysphagia suggests the presence of esophagitis often with an associated stricture.

  Management of scleroderma centers around the treatment of GERD and its complications. Patients should chew their food thoroughly and drink plenty of liquids. Aggressive treatment of heartburn symptoms should be with proton pump inhibitors and prokinetic agents. Strictures respond to frequent dilation. In patients who are not controlled by maximal medical therapy, a modified antireflux procedure can be done, as long as the fundoplication wrap is incomplete.
- Other esophageal motility disorders. The remainder of the esophageal motility disorders include diffuse esophageal spasm (DES), nutcracker esophagus and an ineffective esophageal motility pattern (formerly referred to as a nonspecific motility disorder). The classification of esophageal motility disorders is presented in Table 5.7. Of these the only one which

may be a true "motility" disorder is DES, and that is currently controversial. Most of these are caused by GERD. Endoscopy is helpful in these cases if there is endoscopic or microscopic esophagitis. Occasionally strong contractions can be seen that may give a clue of a motility problem. If any of these are suspected or diagnosed empiric therapy with proton pump inhibitors should be tried. Calcium channel blockers and nitrates have been tried in patients with chest pain or dysphagia ascribed to a motility problem, with varying results. Should the above measures fail, a trial of imipramine (25-50 mg daily) has been proven to be beneficial in patients with noncardiac chest pain.

- **Benign tumors** of the esophagus
  - Intramural-extramucosal tumors. **Leiomyoma** is the most common benign esophageal tumor accounting for two-thirds of benign esophageal tumors. It is usually seen in the distal two-thirds of the esophagus, is usually a single tumor, and is usually round to oval elevations 3-8 cm in diameter. Most leiomyomas are discovered incidentally on endoscopy or UGI barium studies, and due to their intramural location are usually covered by normal epithelium. By endoscopic ultrasound, a leiomyoma appears as a hypoechoic mass arising from the muscularis mucosa (fourth hypoechoic layer). In terms of distinguishing leiomyomas from leiomyosarcoma, EUS, based on size, shape, and appearance is not sensitive to distinguish between these two. Therefore a fine needle aspirate at the time of EUS needs to be considered. Treatment should be based on the patient's symptoms. An asymptomatic patient can be followed clinically, as the chance of malignant transformation is low. If a patient is symptomatic, usually they have retrosternal chest pain or dysphagia, and surgical enucleation is the treatment of choice.
  - Intraluminal tumors. **Lipomas** can occur anywhere in the gastrointestinal tract. They are extremely rare in the esophagus. They usually originate in the submucosa, and protrude into the lumen and can develop a stalk as they grow in size. When they do occur, the most common site is the cervical esophagus. Fibrovascular polyps are found in the upper esophagus and are usually asymptomatic. Most cases are described in elderly men. They are slow growing but can attain large sizes. Endoscopically, these are usually seen on a stalk One concern given their large size and upper esophageal location, is reports of the polyp being regurgitated into the hypopharynx and obstructing the airway, leading to sudden death. The treatment for this polyp is surgical excision.
- **Caustic injury** to the esophagus. Early endoscopy is important in the management of alkali or acid ingestion in order to assess injury and to ascertain when feeding may be started. Caustic ingestion can result in a range of injuries from mild oral burns to sore throat, to esophageal or gastric perforation. Most caustic ingestions occur in the pediatric population. The most common ingested substances are alkaline, such as bleaches, detergents and common cleaning substances (i.e., oven cleaner). Injury from alkali takes the form of a burn and can be graded from first degree burns with mild mucosal sloughing to third degree burns with full thickness injury. Acid injury is less common. Injury occurs as coagulation necrosis with the formation of an eschar. Endoscopy is most important in assessing caustic

trauma and should be performed within 24 hours of the ingestion in order to evaluate the esophagus prior to stricture formation. If minimal injury has occurred, the patient can be started on liquids and sent home with follow-up barium swallow at 3 weeks. In patients with more severe injury, oral nutrition can be started once the patient can swallow. TPN may be necessary along with possible intravenous corticosteroids in patients with severe burns. The risk associated with severe burns is that they can develop strictures that may need repeated dilation.

- Management of esophageal foreign bodies. **Foreign bodies** lodged in the esophagus are seen commonly in practice. In each situation, the type of ingested substance, the patient's condition and the symptoms determine management. Patients who present with acute esophageal obstruction may relate a history of intermittent dysphagia, choking while eating, or of food lodging in their chest. Inability to tolerate secretions is an ominous sign and necessitates prompt endoscopy. Food, in particular meat, is the most common object that is encountered. Typically food can lodge just proximal to a stricture or the GE junction. If the food is impacted for more than 12 hours endoscopy is indicated. If the foreign body passes spontaneously, an elective EGD can be performed. Techniques to remove the foreign body include removing the intact bolus with a polypectomy snare, a polypectomy basket, or the suction cylinder of the variceal band ligator. Use of an overtube may facilitate removal as well as protect the airway. Boluses of a small size can be gently pushed into the stomach. If a foreign body is lodged at the level of the UES, an otolaryngologist should attempt removal with a laryngoscope. Care must be taken with sharp objects and batteries lodged in the esophagus. Using an overtube to protect the airway, a battery can be retrieved using a Roth retrieval net, and objects can be removed with the Roth net or rat-tooth forceps or a snare. If food impaction is caused by a bone, a plain film should be done to identify the site of impaction and to rule out perforation.

5

Selected References

1. Boyce GA, Boyce HW. Esophagus: Anatomy and structural anomalies. In: Yamada T, ed. Textbook of Gastroenterology. 3rd ed. Lippincott Williams and Wilkins: Philadelphia, 1999.
2. Devault KR, Castell DO. Guidelines for the diagnosis and treatment of gastroesophageal reflux disease. Arch Int Med 1995; 155:2165-2173.
3. Faigel DO, Fennerty MB. Miscellaneous diseases of the esophagus. In: Yamada T, ed. Textbook of Gastroenterology. 3rd ed. Lippincott Williams and Wilkins: Philadelphia, 1999.
4. Heudebert GR, Marks R, Wilcox CM et al. Choice of long-term strategy for the management of patients with severe esophagitis: A cost utility analysis. Gastroenterology 1997; 112:1078-1086.
5. Kahrilas PJ. Motility disorders of the esophagus. In: Yamada T, ed. Textbook of Gastroenterology. 3rd ed. Lippincott Williams and Wilkins: Philadelphia, 1999.
6. Marks RD, Richter JE. Peptic strictures of the esophagus. Am J Gastroenterol 1993; 88:1160-1173.
7. Orlando RC. Reflux esophagitis. In: Yamada T, ed. Textbook of Gastroenterology. 3rd ed. Lippincott Williams and Wilkins: Philadelphia, 1999.
8. Richter JE. Editorial comment: Menu for successful antireflux surgery. Dig Dis 1996; 14:129-131.

9. Vigneri S, Termini R, Leandro G et al. A comparison of five maintenance therapies for reflux esophagitis. N Engl J Med 1995; 333:1106-1110.
10. Waring JP, Hunter JG, Oddsdottir M et al. The preoperative evaluation of patients considered for laparoscopic antireflux surgery. Am J Gastroenterol 1995. 90:35-38.
11. Wilcox CM. Esophageal infections, including disorders associated with AIDS. In: Yamada T, ed. Textbook of Gastroenterology. 3rd ed. Lippincott Williams and Wilkins: Philadelphia, 1999.

CHAPTER 6

Malignant Esophagus

Steven J. Shields

Introduction

Malignant tumors of the esophagus are primarily a disease of adults, typically occurring in the sixth or seventh decade. Squamous cell carcinoma and adenocarcinoma account for over 95% of malignant tumors of the esophagus (Table 6.1). Despite well-established epiddemiologic differences beetween the two, survival and mortality rates are equivalent at similar disease stages. Identification of esophageal tumors at an early disease stage has the greatest impact on survival. However, there is no screening program in the U.S. that has been proven effective in high risk patients.

Incidence/Epidemiology

Worldwide

- Tremendous variation in incidence between countries and within countries.
  China 31.49 in 100,000. Middle Africa 2.37 in 100,000
  Linxian, China 700 in 100,000; Fanxian, China <20 in 100,000
- Higher incidence in developing vs developed countries
- 8th most common cancer worldwide
- 6th most common caus eof death from cancer worldwide
- Male to female ration nearly equal in high incidence regions
- Male to female ration significantly higher in low incidence regions

United States

- Significant variation in incidence between men and women
  Males 5.20 in 100,00; (13th out 25 world regions); females 1.41 in 100,000; (16th out of 25 world regions)
- Significant variation in incidence between ethnic groups
  Causcasian males 5 in 100,000; Black males 16.8 in 100,000
- Significant variation in incidence within regions
  Black males 16.8 in 100,000; Black males in Washington, DC region 28.6 in 100,000
- Estimated new cases of esophageal cancer in 1999
  Males 9400; females 3100

Gastrointestinal Endoscopy, edited by Jacques Van Dam and Richard C. K. Wong.

Table 6.1 Malignant tumors of the esophagus

Squamous cell carcinoma
Adenocarcinoma
Lymphoma
Leiomyosarcoma
Leiomyoblastoma
Small cell carcinoma
Metastatic breast cancer
Metastatic lung cancer
Metastatic melanoma
Kaposi's sarcoma

Squamous Cell Carcinoma vs Adenocarcinoma of the Esophagus

- Squamous cell carcinoma used to account for 90-95% of malignant tumors of the esophagus.
- Adenocarcinoma of the esophagus now accounts for 50% of new cases.
- The increasing proportion of adenocarcinoma of the esophagus has been occurring over the past 20 years.
- Adenocarcinoma of the esophagus has been rising at a greater rate of acceleration than any other cancer in the U.S.
- Squamous cell carcinoma occurs in blacks maore than in whites by a 6:1 ratio.
- Adenocarcinoma occurs in whites more than in blacks by a ratio of 4:1.
- Squamous cell carcinoma affects the proximal and middle esophagus primarily; adenocarcinoma affects the distal esophagus primarily.

Etiology/Risk Factors

Squamous cell carcinoma of esophagus. There are numerous risk factors and chronic conditions predisposing to the development of esophageal squamous cell carcinoma. In the U.S. the main risk factors are exposure ot alcohol and tobacco. Worldwide, squamous cell carcinoma has been linked to dietary exposures, vitamin or mineral deficiencies and improper storage of cereals and grains (Table 6.2).

Adenocarcinoma of esophagus. The only clear association with adenocarcinoma of the esophagus is Barrett's esophagus. Conditions predisposong to chronic acid reflux (including nicotine, obesity and certain medications) may be direct risks for development of esophageal adenocarcinoma.

Clinical Presentation

Early stage tumors are typically asymptomatic. Advanced tumors present with a variety of symptoms.

- Dysphagia (solids > liquids) (90%)
- Weight loss (70%)
- Chest pain (60%)
- Weakness (50%)
- Lymphadenopathy (20%)
- Hoarse voice (15%)
- Cough, aspiration pneumonia (10%)
- GI bleeding/anemia (7%)

Table 6.2 Etiology of esophageal squamous cell carcinoma

Dietary and nutritional risk factors
Alcohol
Tobacco
Nitrosamines
Betel nut chewing; animal hide chewing
Tannins (coffee, tea)
Hot tea
Improper storage of cereals and grains
Chronic predisposing conditions
Achalasia (16-fold increase; 340 in 100,000)
Chronic esophagitis
Plummer-Vinson syndrome
Tylosis
History of head and neck cancer
History of partial gastrectomy
Celiac sprue
Chronic infection with human papilloma virus of fungi

Laboratory Evaluation

- Hypoalbuminemia (malnutrition, weight loss)
- Iron deficiency anemia (tumor ulceration and bleeding, Plummer-Vinson syndrome)
- Elevated liver function tests (LFTs) (liver metastases)
- Hypercalcemia (bone metastases, tumor humoral factor, occurs in up to 25% of patients)

Diagnostic Evaluation

- Chest x-ray
  - Air-fluid level in esophagus (high-grade luminal obstruction)
  - Mediastinal widening
  - Tracheal deviation
    - Retrotracheal stripe thickening
    - Lung metastases
- Barium esophagram
  - Location of obstructing lesion (often misses non-obstructing, early-stage lesions)
  - Length of tumor
  - Identification of esophago-respiratory fistula (ERF)
  - Avoid gastrograffin if obstruction or fistula is suspected as aspiration could lead to pulmonary edema
  - Unable to provide histologic or cytologic diagnosis
- Upper endoscopy
  - Location of obstructing lesion
  - Length of tumor
  - Identification of ERF
  - Endoscopic biopsy and cytologic specimens
  - Therapeutic intervention (endoscopic mucosal resection for early, superficial tumors; endoscopic ultrasound probes [EUS]; esophageal dilation [ballonn, Savory dilator]; esophageal stent prosthesis; laser; photodynamic therapy [PDT])

Table 6.3 TNM staging classification of esophageal carcinoma

Primary Tumor	Regional Lymph Nodes (N)*	Distant Mets
TX: Primary tumor can't be assessed	NX: Regional nodes can't be assessed	MX: Distant mets can't be assessed
TO: No evidence primary tumor	NO: No regional node mets	M0: No distant of mets
Tis: Carcinoma in situ (high grade dysplasia)	N1: Regional node mets	M1: Distant mets
T1m: Tumor confined to mucosa		
T1sm: Tumor invading submucosa		
T2: Tumor invading muscularis propria		
T3: Tumor invading adventitia		
T4: Tumor invading adjacent structures		

*For tumors involving the cervical esophagus, the cervical and supraclavicular nodes are considered regional; for tumors involving the intrathoracic esophagus, the mediastinal and perigastric nodes are considered regional.

6

- Balloon cytology
  - Balloon covered by mesh for cytologic study of esophageal cells
  - Endoscopy not required
  - Mass screening in China; 90% accurate in cancer detection; 75% of tumors were in situ or minimally invasive
  - Poor accuracy in low incidence areas of the world
  - Poor distinction between acute inflammation and dysplasia

Staging of Esophageal Tumors

- TNM staging system (Table 6.3)
  - Depth of primary tumor invasion (T)
  - Regional lymph node disease (N)
  - Distant metastasis (M) to solid organs or lymph nodes outside a specificed regional lymph node
  - Prognosis and survival are directly related to tumor stage.
  - Curative vs palliative surgery is directly related to tumor stage
  - Metastatic and advanced tumor stage (T4) are usually considered nonoperative candidates

Table 6.4 Tumor stage

Stage 0	Tis N0M0
Stage I	T1 N0M0
Stage IIA	T2 N0M0
	T3 N0M0
Stage IIB	T1 N1M0
	T2 N1M0
Stage III	T3 N1M0
	T4 N0M0
	T4 N1M0
Stage IV	any T, any N, M1

6

- Endoscopic Ultrasound (EUS)
  - Excellent high-resolution images of the layers of the esophageal wall
  - Superior modality for determination of primary tumor (T) stage
  - Superior modality for determination of regional lymph node (N) status
  - EUS-directed fine needle aspiration (FNA) of abnormal lymph nodes preoperatively
  - EUS can document metastatic (M1) disease to the liver or celiac lymph nodes
- Abdominal CT scan
  - Mid-esophageal tumors better demonstrated than proximal or distal tumors
  - Less accurate at detetcing lymph node metastases (N1) than EUS
  - Better at detecting subdiaphragmatic lymph nodes than mediastinal
  - Excellent at detecting distant metastases (M1), especially to solid organs
- Laparoscopy
  - May detect small liver and intraperitoneal metastases not detectable by other imaging modalities
  - Preoperative evaluation prior to esophageal resection
  - Increase preoperative staging accuracy
  - Added morbidity and cost
- Thoracoscopy
  - May detect mediastinal lymph node metastases
  - Increase preoperative staging accuracy
  - Added morbidity and cost
- Bronchoscopy
  - May allow detection of airway invasion
  - Identification of tumors with increased risk of ERF during radiation and/or chemotherapy
  - Should be performed in advanced tumors located in the proximal esophagus

Treatment of Esophageal Cancer

- Surgical
  - Combined laparotomy and thoracotomy with intrathoracic anastomosis of stomach and esophagus
  - Extrathoracic approach with anastomosis of the stomach to the cervical esophagus
  - Esophagectomy with intestinal interposition
  - Extensive en bloc lymph node resection

Table 6.5 Complications of esophageal endoprostheses

Peforation
Migration
Over/ingrowth
Food impaction
Hemorrhage
Procedural mortality
Mortality (30 day)

Pre- vs postoperative adjuvant chemoradiation
High postop morbidity (up to 75%)
Operative mortality <10% in experienced centers
Most surgery in U.S. is palliative

- Radiation therapy
  External beam radiation
  Intraluminal radiation (brachytherapy)
  "Curative" by itself on occasion (squamous cell >adenocarcinoma)
  Best responses in women, proximal tumors, and tumors <5 cm in length
  Complications include esophagitis, stricture, pneumonitis, pulmonary fibrosis, pericardial effusion, transverse myelitis, ERF
- Chemotherapy
  Cisplatin
  5-fluorouracil (5-FU)
  Paclitaxel
  Ineffective in achieving local tumor control or improving survival
  May have some benefit in metastatic disease control
- Combined chemoradiation
  Chemotherapy seems to potentiate the effects of radiation therapy
  Improved local disease control
  Dose and timing of each not established
  Morbidity increases with higher dose of either
  Severe side-effects in up to 40%
  Life-threatening side effects in up to 20%
  Improved survival if complete or partial responders (as compared to non-responders)

Endoscopic Management

- Endoscopic mucosectomy
  Limited to esophageal tumors involving the mucosa layer only
  Submucosal injection/snare resection technique
  Suction device/snare
  Curative resection in the majority of cases; 92.5% 5 yr survival
  Perforation rate 2-3%
- Photdynamic therapy (PDT)
  Intravenous administration of light-sensitive porphyrin derivative
  Accumulates preferentially in tumor cells
  Activation by low energy laser delivered via an endoscopic probe
  Photochemical reaction leads to cytotoxicity and tumor necrosis

6

Table 6.6 Treatment options for esophageal cancer

Treatment	Advantages	Disadvantages
Surgical resection	Potentially curative	High morbidity, high cost; low cure rate
Radiation therapy	Potential complete response; Improves local tumor control	Moderate morbidity; high cost; multiple treatments; 4-6 wk delay in relief of dysphagia; stricture formation (to 50%)
Chemotherapy	Potential impact on mets	Low complete response rate
Chemoradiation	Complete response improves survival	High morbidity; potential life threatening complications in up to 20%; high cost
Photodynamic therapy (PDT)	Selective tumor ablation; May be curative in mucosal tumors	Skin photosensitivity (6 wk)
Laser therapy	Excellent palliation of dysphagia	High equipment costs
Thermal ablation	Inexpensive	Difficult to control
Sclerosant	Inexpensive	Difficult to control
Endoscopic injection	Immediate palliation; easy	Brief duration
Dilation endoprosthesis	Effective palliation of dysphagia; effective for airway fistula	No impact on survival; metal stents expensive

6

Table 6.7 Esophageal carcinoma: survival

Tumor Stage	5-year Survival
Stage I	60%
Stage II	30%
Stage III	20%
Stage IV	40%

Better patient toelrance and fewer perforations when compared to standard laser therapy
Transient (6 wk) skin photosensitivity (sevre sunburn reactions)
Excellent palliation of dysphagia
Used as curative therapy in early stage disease
New sensitizers (ALA) may reduce skin photosensitivity

Table 6.8 Esophageal carcinoma: survival

TNM Staging	5-year Survival
Tis	80%
T1	46%
T2	30%
T3	22%
T4	7%
N0	40%
N1	17%
M1	3%

- Endoscopic laser therapy
  - Nd:YAG (neodymium:yttrium-aluminum-garnet) laser
  - Thermal ablation via endoscopic targeting of tumor tissue
  - Best results for short, exophytic, noncircumferential tumors in the mid or distal esophagus
  - Retrograde treatment preferable
  - Excellent palliation of dysphagia
  - Perforation rate 2%
- Endoscopic injection therapy
  - Direct sclerosant injection into the tumor under endoscopic guidance
  - Most commonly used sclerosant is absolute alcohol
  - Easy, simple, inexpensive
  - Complications can occur if sclerosant tracks into normal tissue
- Bipolar electrocoagulation
  - Thermal ablation delivered circumferentially from an electrocautery probe passed into the lumen of the tumor under endsocopic guidance
  - Best suited for circumferential, exophytic tumors
  - Effective palliation of dysphagia
- Endoscopic dilation
  - Passage of tapered dilating catheters over a guidewire
  - Passage of dilatin balloon catheter
  - Immediate palliation of dysphagia
  - Often necessary prior to other endoscopic interventions
- Esophageal endoprostheses
  - Insertion of a plastic or metallic stent to maintain a patent esophageal lumen
  - Excellent, immediate palliation of dysphagia
  - Treatment of choice for tumors associated with ERF
  - Increased complications with chemoradiation therapy

Complications: plastic vs. metallic stents (Table 4)

Esophago-Respiratory Fistula (ERF)

- Connection between esophagus and airway
- May occur spontaneously or after treatment of esophageal tumors
- Highest incidence in advance stage tumors of cervical esophagus
- 90% are symptomatic (recurrent cough when swallowing, aspiration, pneumonia, fever, dysphagia)

- 10% are asymptomatic
- Extremely poor prognosis, especially if delay in diagnosis and treatment
- Treatment is palliative only. endoprosthesis is treatment of choice; surgical resection; chemotherapy; radiation therapy

Survival

- Tumor stage and survival are directly related
- Surgical resection offers the potential for cure
- Most surgery in the U.S. palliative
- Combination chemotherapy and radiation therapy with surgery may offer survival advantages, especially in those with complete response
- Screening programs for high-risk patients have not impacted survival in the U.S.

Selected References

1. Lightdale CJ. Esophageal cancer: Practice guidelines. Am J Gastroenterol 1999; 94:20-29.
   Excellent discussion on the staging and treatment of esophageal carcinoma.
2. Wiersema MJ, Vilmann P, Giovanni M et al. Endosonography-guided fine needle aspiration biopsy: Diagnostic accuracy and complication assessment. Gastroenterology 1997; 112:1087-1095.
   EUS-guided FNA is a safe and effective technique.
3. Bosset JF, Gignoux M, Triboulet JP et al. Chemoradiotherapy followed by surgery compared with surgery alone in squamous cell cancer of the esophagus. N Engl J Med 1997; 337:161-167.
   No difference in survival between the two treatment groups.
4. Walsh TN, Noonan M, Hollywood D et al. A comparison of multimodel therapy and surgery for esophageal adenocarcinoma. New Engl J Med 1996; 335:462-467.
5. Herskovic A, Mratz K, al-Sarraf M et al. Combined chemotherapy and radiotherapy compared to radiotherapy alone in patients with cancer of the esophagus. New Engl J Med 1992; 326:1593-1598.
6. Ziegler K, Sanft C, Zeitz M et al. Evaluation of endosonography in TN staging of esophageal cancer. Gut 1991; 32:16-20.
   EUS is superior to CT scan for determining TN-stage of esophageal cancer.
7. Landis SH, Murray T, Bolden S et al. Cancer statistics 1999. CA Cancer J Clin 1999; 49:8-64.
   Current statistic of esophageal cancer in U.S. and throughout the world, including incidence by gender and ethnicity.
8. Lightdale CJ, Heier SK, Marcon NE et al. Photodynamic therapy with porfimer sodium versus thermal ablation with Nd:YAG laser for palliation of esophageal cancer: A multicenter randomized trial. Gastrointest Endosc 1995; 42:507-612.
9. Mellow MH, Pinkas H. Endoscopic laser therapy for malignancies affecting the esophagus and gastroesophageal junction: Analysis of technical and functional efficacy. Arch Intern Med 1985; 145:1443-1446.
10. Gevers AM, Macken E, Hiele M et al. A comparison of laser therapy, plastic stents, and expandable metal stents for palliation of malignant dysphagia in patients without a fistula. Gastrointest Endosc 1998; 48:382-388.
11. Siersema PD, Hop WCJ, Dees J et al. Coated self-expanding metal stents versus latex prostheses for esophagogastric cancer with special reference to prior radiation and chemotherapy: a controlled, prospective study. Gastrointest Endosc 1998; 47:13-120.

12. Dumonceau JM, Cremer M, Lalmand B et al. Esophageal fistula sealing: choice of stent, practical management and cost. Gastrointest Endosc 1999; 49:70-79.
13. Nelson DB, Axelrad AM, Fleischer DE et al. Silicone-covered Wallstent prototypes for palliation of malignant esophageal obstruction and digestive-respiratory syndromes. Gastrointest Endosc 1997; 45:31-37.
14. Catalano MF, Sivak MVJ, Rice T et al. Endosnographic features presidctive of lymph node metastasis. Gastrointest Endosc 1994; 40:442-446.

CHAPTER 7

Esophageal Manometry

Brian Jacobson, Nathan Feldman and Francis A. Farraye

Introduction

Esophageal manometry is a procedure in which intraluminal pressures are measured at different sites along the length of the esophagus including the upper and lower esophageal sphincters. The pattern of observed pressures (including the amplitudes, duration and peristaltic properties) provides information about possible diseases affecting the esophagus. Esophageal manometry involves either naso-esophageal or oropharyngeal-esophageal intubation with a specially designed pressure transducer. A manometry study usually lasts 30-40 minutes, depending on the amount of information to be collected. Manometry readings are recorded as continuous pressure tracings from one to eight sites simultaneously (Fig. 7.1).

Relevant Anatomy

- Nares, inferior nasal turbinate, nasal septum. A deviated septum may prevent safe passage of a manometry probe
- Hypopharynx and upper esophageal sphincter (UES). UES is composed of the cricopharyngeus muscle and parts of the inferior pharyngeal constrictor. These are composed of striated muscle fibers. Innervation of the UES is provided by the vagus nerve.
- Esophageal body. The esophageal wall is composed of four layers: the mucosa, the submucosa, the muscularis propria, and the serosa.
- Mucosa inner lining of the esophagus lined by squamous epithelial cells until the "squamo-columnar junction" where the lining is replaced by columnar epithelium. The squamo-columnar junction is also called the "Z-line" because it has a jagged or zigzag appearance. It occurs normally at the level of the lower esophageal sphincter.
- Submucosa comprised of collagen and elastic fibers
- Muscularis propria. Has two parts; an inner circular layer, so called because its muscle fibers are arranged circumferentially around the esophageal lumen. There is also the outer longitudinal layer with its muscle fibers oriented along the long axis of the esophagus. The proximal third of the esophagus contains striated muscle and is innervated by the vagus nerve. The distal two thirds contain smooth muscle and receives parasympathetic innervation from the vagus nerve and sympathetic input from the celiac ganglia and the sympathetic trunk. There is also a complex enteric nervous system between the two layers of the muscularis propria.
- Serosa. Comprised of connective tissue
- Lower esophageal sphincter
  - Thickened ring of muscle 3-5 cm long
  - Innervated by the vagus nerve and the enteric system within the esophagus

Gastrointestinal Endoscopy, edited by Jacques Van Dam and Richard C. K. Wong.

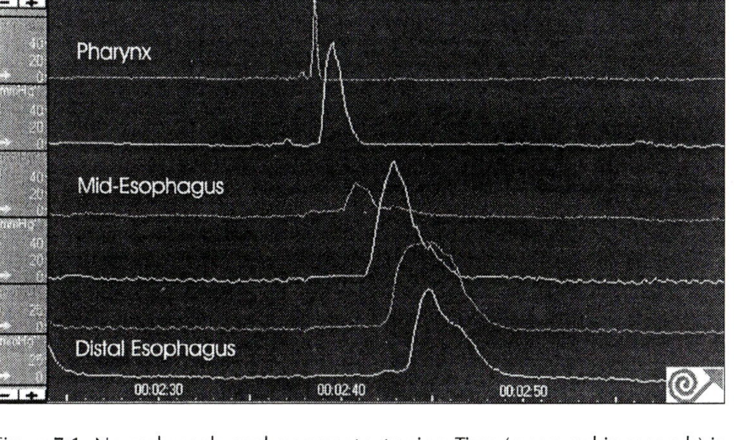

Figure 7.1. Normal esophageal manometry tracing. Time (measured in seconds) is located on the horizontal axis. Pressure (measured in mm Hg) is located on the vertical axis. Each line reflects the pressures measured at a single point along a manometry catheter. The ports are spaced 5 cm apart, except for the very tip where there are two ports 1 cm apart. Each tracing represents a different position along the esophagus. The descending order of tracings represents progressively lower points of the esophagus. In response to a swallow, the amplitude rises and falls at each point along the esophagus with a continuously advancing front from superior to inferior (courtesy of Medtronic Functional Diagnostics, Inc.).

- Cardia and fundus of stomach innervated by the vagus nerve
- The fundus relaxes during deglutition to accommodate a food bolus

Indications and Contraindications

- Indications. To establish the diagnosis of achalasia.
- Achalasia is a disorder of unknown etiology in which there is loss of nerves in the myenteric plexus
- Suspected clinically because of dysphagia to solids and liquids, regurgitation of undigested food (sometimes hours after eating), cough, weight loss, and rarely chest pain. A barium swallow may demonstrate a dilated esophagus with a "bird's beak" appearance of the distal esophagus. Pseudoachalasia, a disorder seen more commonly in the elderly, in which manometric findings are similar to achalasia but are due to a malignancy
- To establish the diagnosis of diffuse esophageal spasm (DES). DES is a rare condition of unknown etiology in which there are simultaneous contractions at various levels of the esophagus suspected clinically because of intermittent substernal chest pain that may be precipitated by a barium swallow which may reveal multiple simultaneous contractions, the so-called "corkscrew" esophagus
- To detect esophageal motor abnormalities associated with connective tissue diseases (such as scleroderma). The American Gastroenterological Association recommends that esophageal manometry be performed only if detecting an abnormality will aid in diagnosing a systemic disease or if it will affect patient management.[1]

- To identify the location of the lower esophageal sphincter for accurate placement of a pH probe (see chapter on 24-hour pH monitoring).
- May provide a preoperative assessment of peristaltic function prior to anti-reflux surgery
- Should NOT be used to diagnose gastroesophageal reflux disease (GERD). Poor esophageal contractile function MAY correlate with GERD, but its presence will not make the diagnosis. 24-hour pH monitoring may be more useful for this indication (see chapter on 24-hour pH monitoring).
- NOT appropriate for the initial evaluation of nonspecific chest pain or esophageal symptoms. Provocative tests that do not involve manometry can be used to reproduce a patient's pain syndrome. The **edrophonium** or **Tensilon test** induces motor abnormalities by increasing acetylcholine activity in the esophagus. The **Bernstein test** consists of the instillation of a small amount of hydrochloric acid into the esophagus. The balloon-distension test uses a small balloon which is expanded in the esophagus. These tests determine esophageal sensitivity more than motor activity.
- Contraindications. Presence of an esophageal obstruction and risk of perforation.
  - Presence of a large esophageal diverticulum

    Should be suspected if patient regurgitates undigested food hours after eating. Diverticula are best diagnosed by barium swallow when suspected.

Equipment and Accessories

- Manometric apparatus. Pressure sensor and transducer: detects esophageal pressure and converts it to an electrical signal. Two design types: Water-perfused manometric catheters which require a pneumohydraulic pump and volume displacement and solid state system with strain gauges.
- Transducing device to convert pressure readings into electrical signals (Fig. 7.2)
- Miscellaneous accessories for nasal intubation:
  - topical anesthetic;
  - water-soluble lubricant;
  - emesis basin.
- Computer for storage and analysis of data
- Recently portable units have been designed for use during long-term, ambulatory monitoring. These can be combined with pH monitoring for precise measurements of both esophageal pH changes and motor function.[2]

Technique

- Patient preparation
  - npo after midnight
  - Informed consent must be obtained
  - Positioning: best done with the patient sitting
  - Anesthesia: a topical spray into the nose
  - passage of probe: similar to the passage of a nasogastric tube
- **Lower esophageal sphincter (LES) pressure**
  - All recording sites located in the stomach. Confirm the location by recording a positive deflection with breathing or abdominal pressure. The LES pressure is measured by withdrawing the catheter at a rate of 1cm/sec, during a breath hold. The catheter is withdrawn multiple times and multiple recordings are made.

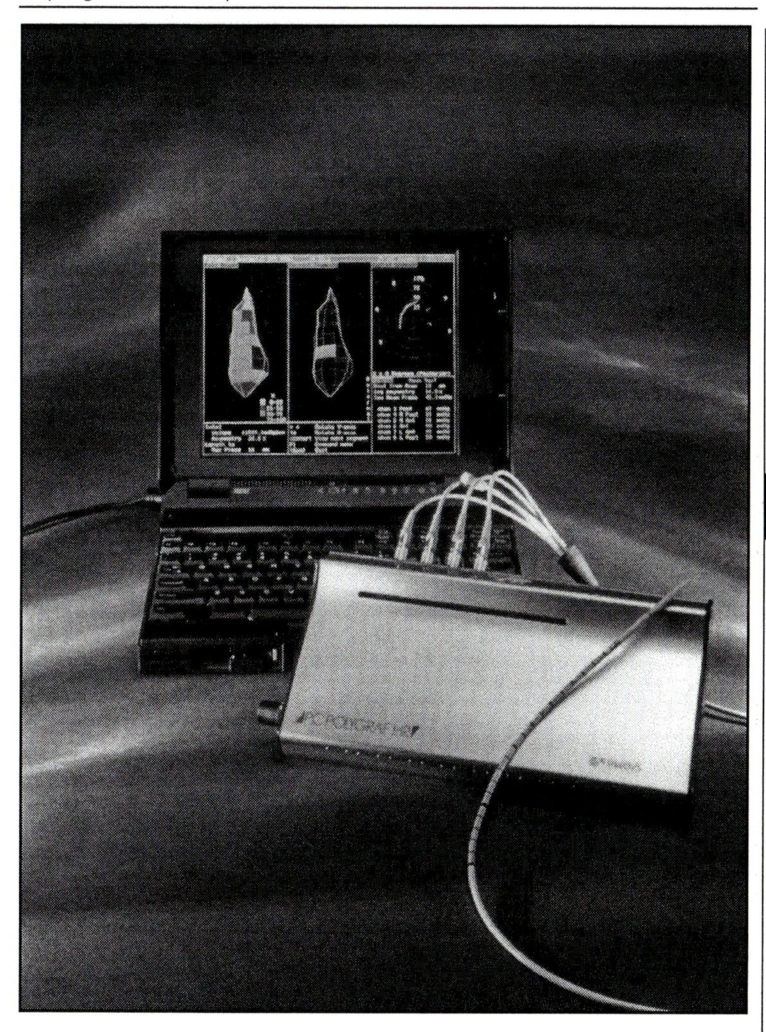

Figure 7.2. A mutichannel transducer and a manometry catheter.

- The slow pull-through or stationary technique. In this case the catheter is withdrawn in 0.5 to 1 cm increments, leaving it in position to measure both peak pressure and relaxation at each level of the probe. Normal value ranges from 15-40 mm Hg. LES relaxation is also measured. In this case, the pressure should fall appropriately at the ONSET of a swallow and remain relaxed until peristalsis travels down the entire esophageal body (Fig. 7.3).

- **Esophageal body pressures**: All recording sites are withdrawn into the esophagus. The patient is given water to drink and a series of wet swallows are used to obtain pressure recordings from the distal esophagus. The catheter is withdrawn

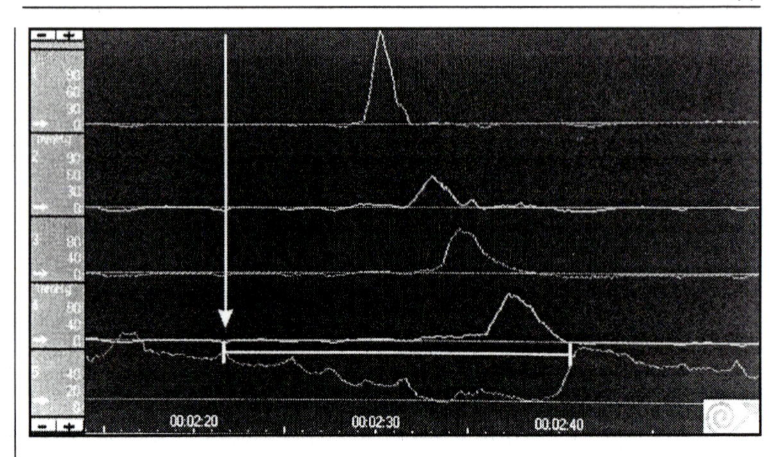

Figure 7.3. Normal lower esophageal sphincter (LES) relaxation. Note the fall in LES pressure that accompanies a swallow. Failure of the LES to relax appropriately can be found in esophageal motility disorders such as Achalasia (courtesy of Medtronic Functional Diagnostics, Inc.).

in 3-5cm intervals and additional recordings are made, allowing capture of the pressures in the proximal esophagus. Dry swallows may also be measured. Generally wet swallows have higher amplitudes than dry swallows. The range of normal pressures for the esophageal body varies from 50 to 100 mmHg. Note is also made of the duration of contractions (normal being 3-4 seconds) and the coordinated movement of peristalsis (contractions progressing in an orderly fashion from proximal to distal esophagus).

- **Upper esophageal sphincter (UES) pressure**: measured using a technique similar to that used to record LES pressure. The recording speed may need to be increased from 2.5 mm/sec to 5-10 mm/sec. The UES pressure is often difficult to measure accurately. Motor abnormalities of the UES are better evaluated with videofluoroscopy.

Outcomes

- In a review of their experience with 268 patients referred for esophageal manometry, Johnston et al found that manometry was normal in half of the cases.[3] However, a specific diagnosis was made almost twice as often when the clinical symptom of dysphagia was present. The frequency of specific diagnoses made is shown in Table 7.1. The manometric study altered patient management in half of the cases.
- The essential measurements in esophageal manometry are the magnitude of the contraction, both within the esophageal body and the LES, to determine if the LES relaxes appropriately or is tonically contracted, the presence or absence of peristalsis, and whether peristalsis is orderly or disorderly.[4] These basic measurements are then placed in the clinical context of a patient's presentation. Together, a diagnosis is suggested. There are a limited number of diseases in which

Table 7.1. The frequency of specific diagnoses made after esophageal manometry in a study by Johnston et al[3]

	Number of Patients (%)
Normal	132 (49.3)
Achalasia	48 (17.9)
Diffuse Esophageal Spasm	36 (13.4)
Nutcracker esophagus	7 (2.6)
Hypertensive LES	12 (4.5)
Connective Tissue Disease	21 (7.8)
Other	12 (4.5)
Total	268

the patterns observed in esophageal manometry are helpful. These diseases and their patterns of findings follow.

- Achalasia (Fig. 7.4a, 7.4b)
  - Absence of esophageal peristalsis (hallmark finding)
  - Elevation of LES resting pressure above 45 mm Hg
  - Failure of LES to relax
  - Elevated intraesophageal pressure when compared with gastric pressure
- Diffuse esophageal spasm (Fig. 7.5)
  - Contractions detected at various levels of the esophagus simultaneously
  - Simultaneous contractions occur after more than 20% of wet swallows
  - Contractions may occur spontaneously and as multiple consecutive waves
  - Pressure tracings may show notched peaks
  - Peak pressures can be low or high, and debate exists regarding the significance of the amplitude of contractions.[5,6]
  - Contractions may exceed the normal duration time of 3-4 sec
- Nutcracker esophagus (Fig. 7.6)
  - Pressure measurements >180 mm Hg
  - Contractions may be prolonged
  - LES pressures may be normal or elevated
- Hypertensive LES
  - LES pressure >45 mm Hg
  - Normal esophageal peristalsis
  - Debate exists whether this is a true primary diagnosis
- Scleroderma/CREST Syndrome (Fig. 7.7)
  - Weak contractions in the lower 2/3 of the esophagus
  - Decreased LES pressures
- Nonspecific esophageal motility disorder
  - Label for manometric abnormalities that do not fit a specific pattern, e.g., a patient with dysphagia or chest pain in whom manometry does not provide a definitive diagnosis
  - Findings can include low amplitude contractions, elevated or diminished LES pressure, prolonged contractions, or non-propagating contractions

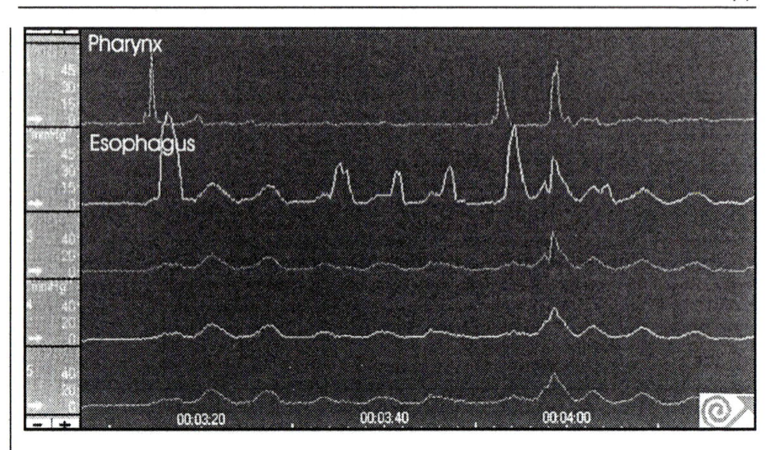

Figure 7.4a. Achalasia recorded from the proximal esophagus. Note that the first swallow, marked by a rise in pressure in the pharynx, is not propagated beyond the first esophageal port. The failure of peristalsis is the manometric hallmark of achalasia (courtesy of Medtronic Functional Diagnostics, Inc.).

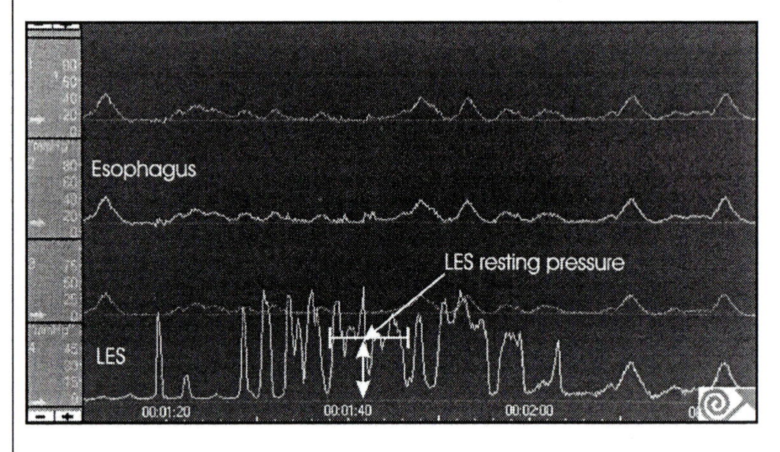

Figure 7.4b. Achalasia measured from the distal esophagus and lower esophageal sphincter (LES). The baseline resting LES pressure is greater than 45 mm Hg. This is a common finding in patients with achalasia (courtesy of Medtronic Functional Diagnostics, Inc.).

Preoperative Evaluation of Patients for Antireflux Surgery

- Helpful in determining the type of surgical procedure to be performed. If manometry demonstrates poor peristalsis, a subtotal fundoplication may yield a lower incidence of postoperative dysphagia compared with a 360° fundoplication. One study found manometry altered surgical decisions in 13% of cases.[7]

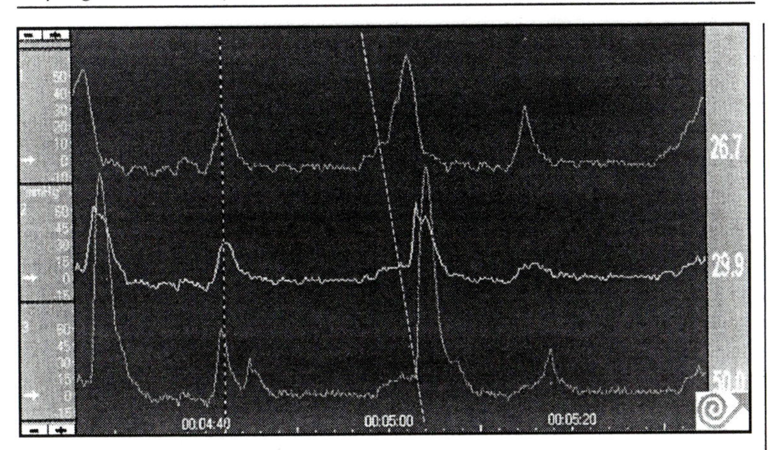

Figure 7.5. Diffuse esophageal spasm (DES). The black and white striped line runs vertically through three pressure peaks, demonstrating simultaneous contractions at various levels in the esophagus. This is the manometric hallmark of DES. Normal peristalsis can be seen in the same tracing (dashed white line) (courtesy of Medtronic Functional Diagnostics, Inc.).

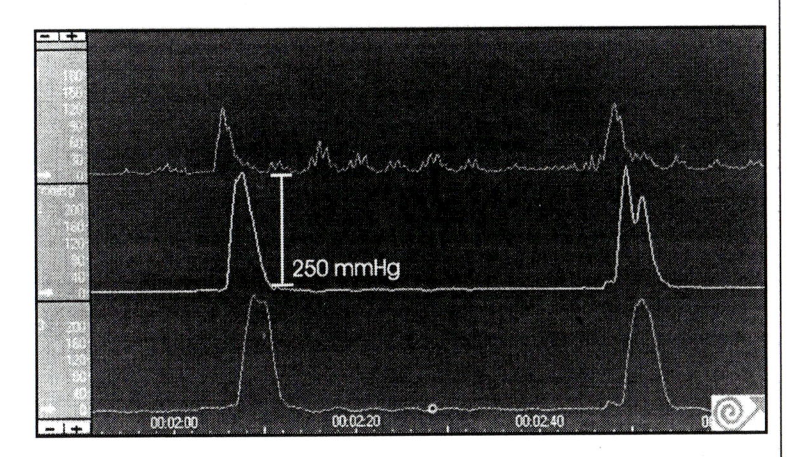

Figure 7.6. Nutcracker esophagus. The characteristic findings here are pressure amplitudes greater than 180 mm Hg and prolonged contractions (lasting over 4 sec) (courtesy of Medtronic Functional Diagnostics, Inc.).

Complications

- In general, a very safe procedure with major complications exceedingly rare. Complications are related to passage of the probe.
- Minor complications: epistaxis, laryngeal trauma, bronchospasm and vomiting.
- Major complications: laryngospasm, pneumonia and esophageal or gastric perforation

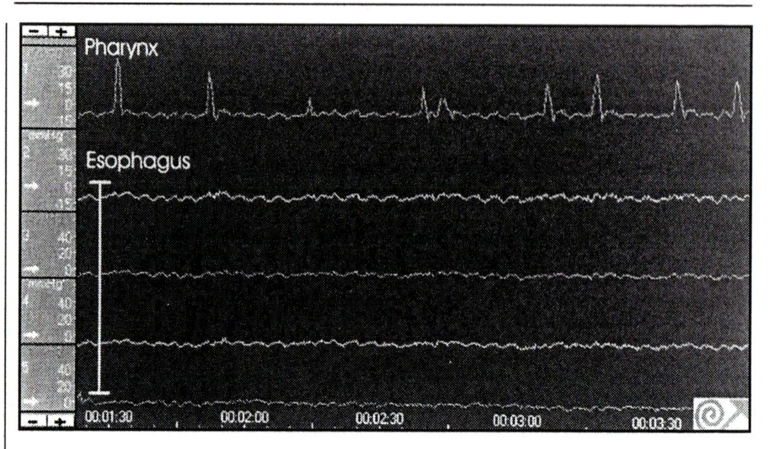

7

Figure 7.7. Manometric tracing from a patient with scleroderma. There is a complete absence of peristalsis (courtesy of Medtronic Functional Diagnostics, Inc.).

Selected References

1. Kahrilas PJ, Clouse RE, Hogan WJ. American Gastroenterological Association technical review on the clinical use of esophageal manometer. Gastroenterology 1994; 107:1865-1884.

   This review presents the official recommendations of the American Gastroenterological Association for the appropriate use of esophageal manometry.

2. Wang H, Beck IT, Paterson WG. Reproducibility and physiologic characteristics of 24-hour ambulatory esophageal manometry/pH-metry. Am J Gastroenterol 1996; 91:492-497.

   The authors performed ambulatory monitoring on two occasions in ten healthy volunteers. A dual-chamber pH probe and a three-channel solid state manometry catheter was used. Intrasubject variability was examined to establish the reproducibility of the technique.

3. Johnston PW, Johnston BT, Collins JSA et al. Audit of the role of oesophageal manometry in clinical practice. Gut 1993; 34:1158-1161.

   The investigators reviewed their experience with 268 patients between 1988 and 1991 at a tertiary care hospital in Belfast. Their center serves a population of 1.5 million people. Most patients had been referred for evaluation of dysphagia, noncardiac chest pain and gastroesophageal reflux disease.

4. Ergun GA, Kahrilas PJ. Clinical applications of esophageal manometry and pH monitoring. Am J Gastroenterol 1996; 91:1077-1089.

   This is a review focusing on the clinical applications of both esophageal manometry and pH monitoring. Attention is given to the role of provocative testing and the emerging technique of ambulatory esophageal manometry.

5. Dalton CB, Castell DO, Hewsone EG et al. Diffuse esophageal spasm. A rare motility disorder not characterized by high amplitude contractions. Dig Dis Sci 1991; 36:1025-1028.

   The authors reviewed their experience with 56 cases of diffuse esophageal spasm. They found only two patients with high-amplitude manometric values. They defined high-amplitude as >180 mm Hg. From a figure in the paper it appears that 12 patients (21.4%) had at least one amplitude greater than 100 mm Hg.

6. Allen ML, DiMarino AJ. Manometric diagnosis of diffuse esophageal spasm. Dig Dis Sci 1996;41:1346-1349.
These investigators studied 60 consecutive patients with manometric criteria of diffuse esophageal spasm. They found two distinct groups: those whose highest manometric amplitude exceeded 100 mm Hg and those whose highest measurement was less than 75 mm Hg. Chest pain was a more commonly reported symptom in those patients with the higher pressure readings. The authors hypothesize that there may be two distinct forms of diffuse esophageal spasm.
7. Waring JP, Hunter JG, Oddsdottir M et al. The preoperative evaluation of patients considered for laparascopic antireflux surgery. Am J Gastroenterol 1995; 90:35-38.
In this prospective study, 88 patients underwent an esophagogastroduodenoscopy (EGD), esophageal manometry, and a 24-hour ambulatory esophageal pH monitoring as part of a preoperative evaluation for antireflux surgery. The role of each procedure in affecting surgical decision making was then reviewed.

CHAPTER 8

Twenty-Four Hour pH Testing

Brian Jacobson, Nathan Feldman and Francis A. Farraye

Introduction

Gastroesophageal reflux disease (GERD) affects millions of people nationwide and accounts for a large proportion of visits to both primary care physicians and gastroenterologists each year. The diagnosis of GERD is often made on clinical grounds. An empirical trial of acid-suppression therapy is a common approach to both bolster the diagnosis and relieve symptoms. In certain patients, esophagogastroduodenoscopy (EGD) may be part of the evaluation and either endoscopic or histologic support for the diagnosis of GERD may be obtained.

However, circumstances arise in which further objective evidence of intra-esophageal acid exposure is needed. Esophageal pH recording allows for measurement of the frequency of acid-reflux episodes and for correlation between reflux events and patient symptoms. Modern pH recording devices are small and lightweight enabling long periods of data acquisition in the ambulatory setting (Fig. 8.1).

Relevant Anatomy (for details see the chapter on esophageal manometry)

- Nares, inferior nasal turbinate, nasal septum
- Hypopharynx and upper esophageal sphincter (UES)
- Esophageal body
- Lower esophageal sphincter
- Cardia, fundus and body of stomach

Indications

Have been divided into true indications and possible indications by the American Gastroenterological Association.[1]

- True indications:
  - To document abnormal esophageal acid exposure prior to antireflux surgery when endoscopy fails to reveal reflux esophagitis.
  - To evaluate the role of acid reflux AFTER antireflux surgery in patients with persistent symptoms.
  - To evaluate patients with GERD symptoms refractory to empiric proton pump inhibitor therapy when endoscopy is normal or equivocal.
- Possible indications:
  - To evaluate noncardiac chest pain.
  - To determine acid reflux in the presence of atypical GERD manifestations (such as asthma, cough, and laryngitis).

Gastrointestinal Endoscopy, edited by Jacques Van Dam and Richard C. K. Wong.

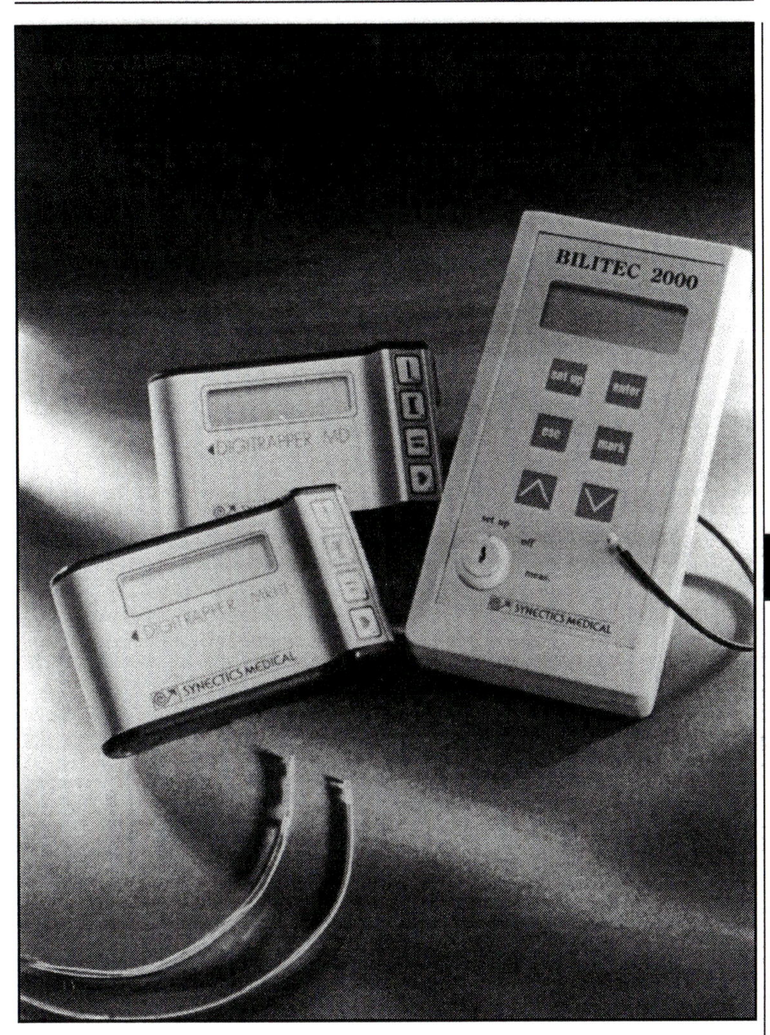

Figure 8.1. These are data loggers used for ambulatory pH monitoring. Also pictured are two pH probes (photograph courtesy of Medtronic Synectics).

Equipment and Accessories

- Data loggers are portable, battery-powered units that store data. Can be attached to a patient's belt or worn with a shoulder-strap. Possess some form of **event recorder** that allows a patient to indicate timing of specific events like the onset of symptoms or the timing of meals. This is later correlated with a written diary kept by the patient.

Table 8.1. The three types of pH electrodes have different characteristics

	Electrodes	ISFET	Glass
Size	Small	Large	Large
Easily Placed	Yes	No	No
Long Life	No	Yes	Yes
Cost	$	$$	$$$
Reference Electrode	Int/Ext	Ext	Int
Sensitivity	+	++	++
Rapid Response To pH Change	+/-	+	+
Multiple Sensor Sites	-	+	-
Used in both Gastric and Esophageal Studies	-	+	+

8

- pH electrodes. There are three basic types of **electrodes** (also called catheters) with different qualities (see Table 8.1).
  - Antimony electrodes
  - Ion-sensitive field effect transistor (ISFET) electrodes
  - Glass electrodes
- The antimony catheters have been the most commonly used. ISFET catheters, however, may become more popular because of several desirable attributes.
- Patient symptom diary
  - A written log on which the patient can record the timing of symptoms and meals (Fig 8.2).
- Miscellaneous accessories for nasal intubation
  - Topical anesthetic
  - Water-soluble lubrication
  - Emesis basin
  - Adhesive tape for affixing the probe to the patient
- Equipment for analysis of data
  - Typically a computer and software that is supplied by the manufacturer of the data logger.
- There are also pH-monitoring devices that detect **bile reflux**. They measure pH values best in the basic, rather than acidic, range. One is pictured in Figure 8.1. The utility of these devices has not been fully evaluated. There are no recommendations by the American Gastroenterological Association for the measurement of alkaline reflux.

Technique

- Patient preparation
  - NPO for at least 6 hours prior to the study.
  - Discontinue all proton pump inhibitors (PPIs) one week prior to the study (except in the evaluation of patients with refractory reflux symptoms and noncardiac chest pain while on PPIs).

Patient:

PATIENT DIARY, 24 HOUR PH MONITORING

CHEST PAIN	REGURGITATION (air,food)	MEALS (Start-Finish)
510 P (LASTED 5 MINUTES)	BELCH WITH LIQUID COMING UP 10:41 AM (THEN COUGHING)	BREAKFAST 7:05 - 7:30 A LUNCH 12:10 - 12:32 SNACK 3:31 P DINNER 6:45 - 7:30 P
HEARTBURN	**NAUSEA**	**SLEEP (Start-Finish)**
11:10 AM 2:43 PM 10:10 PM 2:26 AM		11 PM TO BED 2:26 WOKE UP ~2:30 BACK TO BED 6:30 AM WOKE UP
DIFFICULTY TO SWALLOW	**COUGH**	**MEDICATION**
	COUGHING FIT 10:41 AM (~2 MINUTES)	NONE

Figure 8.2. An example of a symptom diary.

ACID REFLUX		Total	Uprght	Supine	Meal	PostP	HrtBrn	CHPAIN
Duration	(HH:MM)	22:31	13:31	09:00	03:35	06:41	00:04	00:04
Number of reflux episodes	(#)	24	23	1	5	14	1	0
Number of reflux episodes longer than 5.0 minutes	(#)	6	5	1	1	3	0	0
Longest reflux episode	(min)	49	49	24	16	43	4	0
Total time pH below 4.00	(min)	173	149	24	24	111	4	0
Fraction time pH below 4.00	(%)	12.8	18.3	4.5	11.3	27.6	100.0	0.0
Reflux Index	(refl/hour)	1.2	2.1	0.1	1.6	2.9	0.0	0.0
Esophageal Clearance	(min/refl)	6.3	5.3	23.4	4.3	5.7	0.0	0.0
Maximum pH value	(pH)	8.5	8.5	8.2	8.5	7.9	3.2	7.5
Minimum pH value	(pH)	0.8	0.8	2.8	1.8	1.5	2.1	6.3
Symptom Index	(%)	100.0	100.0	100.0	60.0	80.0	100.0	0.0

Numerical scoring on acid refluxes		Scoring Value
Number of reflux episodes	(refl/24 Hours)	1.5
Number of reflux episodes longer than 5.0 minutes	(refl/24 Hours)	5.7
Longest reflux episode	(minutes)	6.4
Fraction time pH below 4.00	Total (%)	9.3
Fraction time pH below 4.00	Uprght (%)	7.8
Fraction time pH below 4.00	Supine (%)	4.9

Total score = 35.6 DeMeester normal values (Adult n = 50)

Figure 8.3. A summary report from a 24-hour pH study. Notice the DeMeester score components are provided as well as a calculated total score (in this case, 35.6).

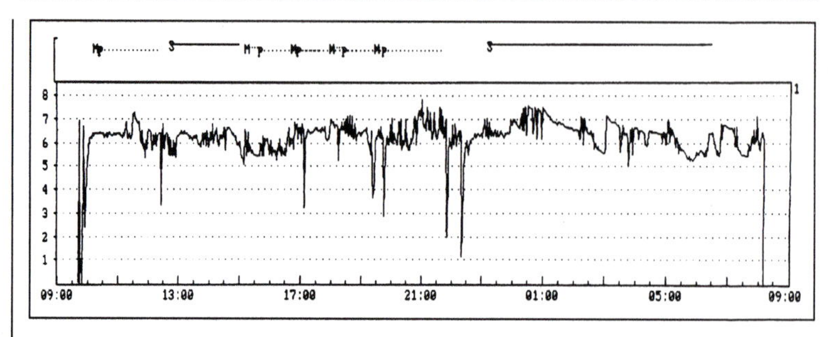

Figure 8.4. A normal pH tracing. Notice the transient dips in pH into the acid range. Overall the majority of time the recording was done, the intraesophageal pH was between 6 and 7. The composite DeMeester score was 2 and the fraction of time that the pH was below 4 was 0.5%. The lines labeled "M", "P", and "S" above the tracing mark meals, post-prandial periods, and supine periods respectively.

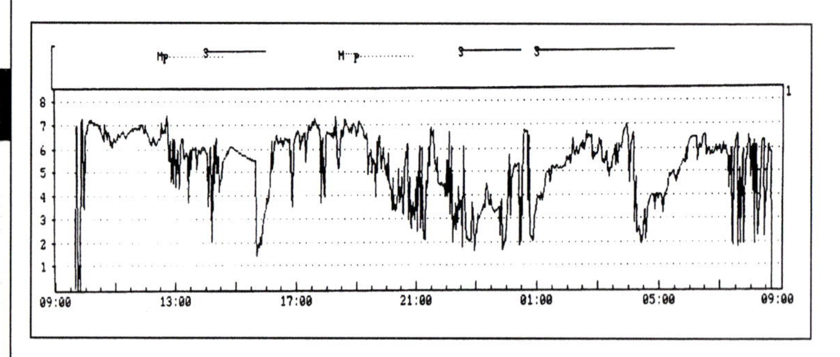

Figure 8.5. A tracing consistent with GERD. This patient's tracing reveals marked fluctuations in pH measurements. During this twenty-four hour period, the pH fell below four 13.6% of the time. The DeMeester score was 56.4.

-Discontinue other antacid medications 24-48 hours prior to the evaluation.
- Obtain informed consent.

- Catheter passage
  - The catheter should be calibrated according to the manufacturer's recommendation.
  - A topical anesthetic is sprayed into the patient's nares.
  - The catheter is lubricated and then passed via the nares into the stomach. The stomach is identified by a drop in measured pH.
  - The catheter is then withdrawn slowly until there is a pH rise, marking the level of the LES.
  - The pH probe is positioned 5 cm above the LES. If a gastric probe is being used, it should be positioned 5 cm below the LES. Some probes have both proximal and distal pH sensors allowing simultaneous measurements of pH in the proximal and distal esophagus.
  - The catheter is taped to the nose and cheeks and is then looped over the patient's ear to hold its position for the duration of the study.

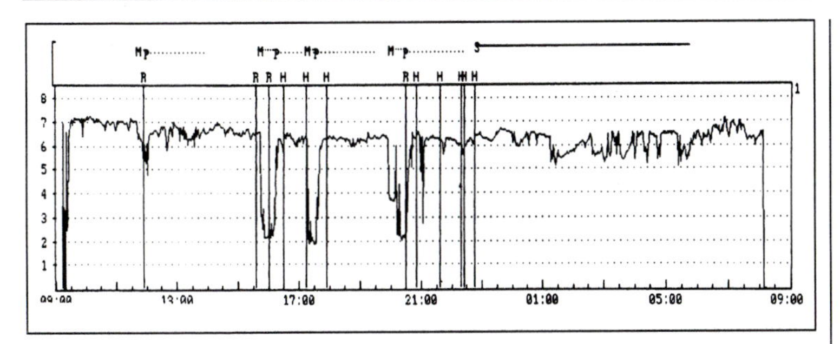

Figure 8.6. An equivocal pH study. The DeMeester score was 13.8 and the fraction of time that the pH was below 4 was 4.2%. While these scores are below the cut-off for the diagnosis of GERD, notice that three episodes of heartburn (H) and regurgitation (R) correlated with drops in pH below 3. One must use clinical judgement in cases such as these to decide if there is "pathological GERD".

- Monitoring period
  - The catheter is left in place for 12-24 hours.
  - The patient keeps a diary of symptoms for the duration of the exam. They can also use event markers that are built into some data-loggers.
  - The patient should avoid showering or bathing while the equipment is in place.
  - There should otherwise be no restrictions on the patient's activities for the duration of the study.
  - The patient does not usually need to restrict their diet. Cigarette smoking may be allowed but should be recorded in the patient's diary.
  - The patient should keep a diary to record the timing of meals, sleep and recumbency, episodes of symptoms, and any other noteworthy events (i.e. vomiting, prolonged coughing fits).

Outcomes

- The original system for scoring 24-hour pH studies was published by Johnson and DeMeester.[2] It looked at six variables:
  - Percentage of total study time pH < 4
  - Percentage of time patient is upright and pH < 4
  - Percentage of time patient is recumbent and pH < 4
  - Number of episodes when pH < 4
  - Number of episodes when pH < 4 for five or more minutes
  - Longest episode when pH < 4 (in minutes)
- Software analyzing the pH study supplies these variables and determines a composite score (Fig. 8.3). Based on the 96th percentile scoring of 50 healthy volunteers, a composite score > 16.7 is consistent with GERD. However, the current consensus is that the only consistently accurate variable is the percentage of total study time that the pH < 44. The upper limit of normal for this value is 4.5% of the total study time.[3]
- The sensitivity of 24-hour pH monitoring in diagnosing GERD ranges from 81 to 96%. The specificity ranges from 85 to 100%.[4]

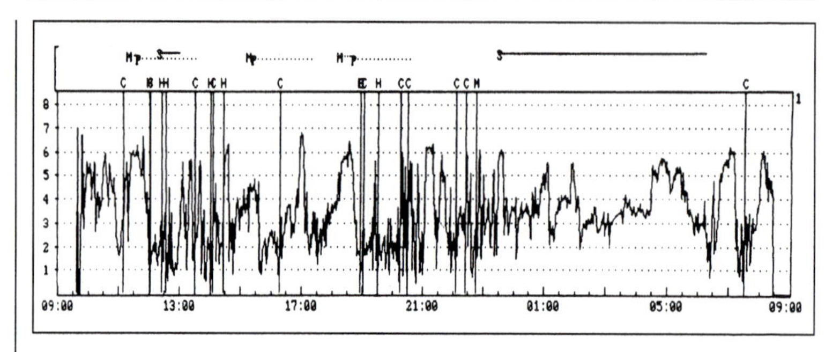

Figure 8.7a. Correlation of symptoms with esophageal acid exposure. Note that the patient indicates chest pain (C) and heartburn (H) at points where the pH is less than 4. However, there are also episodes when the pH is less than 4 where no symptoms are reported. A tracing with this frequency of acid-episodes may indicate that the pH probe slipped into the stomach.

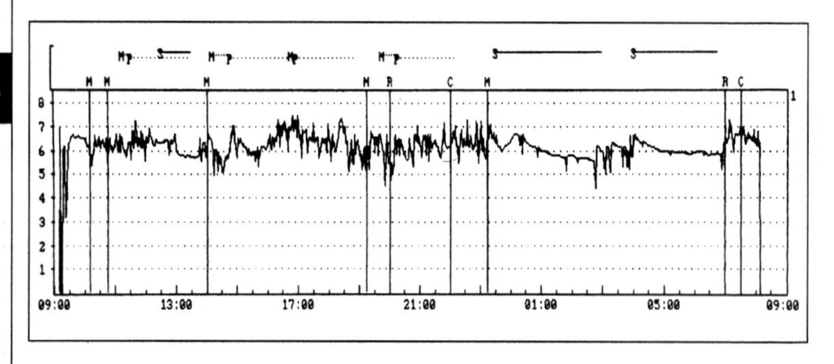

Figure 8.7b. Failure of symptoms to correlate with acid reflux. In this instance, the patient felt chest pain (C) or experienced regurgitation (R) without a significant drop in the measured intraesophageal pH.

- There are three possible outcomes to a pH study:
  - A convincingly negative study (Fig. 8.4).
  - A convincingly positive study (Fig. 8.5).
  - An equivocal study (Fig. 8.6). In this case, correlation (or lack of correlation) of symptoms with episodes of reflux can help decide if the patient has pathological reflux.
- Symptom diaries and tracing marks can be examined for correlation between symptoms and reflux episodes (Fig. 8.7a). However, as Figure 8.7b demonstrates, a correlation may not always exist. Various indices of symptoms and their correlation with esophageal acid-exposure have been developed, but none are consistently reliable.[5]
- The impact of 24-hour ambulatory pH monitoring on preoperative evaluation for antireflux surgery was addressed by Waring et al.[6] Only 38% of the patients

they studied lacked endoscopic features of esophagitis. Of those patients, 85% had an abnormal pH study. Overall only 5% of subjects studied had normal pH studies. None of the patients with normal pH studies underwent antireflux surgery. Therefore, the role of pH monitoring in a preoperative evaluation for antireflux surgery should be limited to those patients whose diagnosis is in doubt and there is no endoscopic evidence for esophagitis.

Complications

- In general this is a very safe study with major complications being exceedingly rare.
- Complications are related to passage of the probe.
- Minor complications
  - Laryngeal trauma
  - Bronchospasm
  - Vomiting
  - Epistaxis
- Major complications
  - Laryngospasm
  - Pneumonia
  - Esophageal or gastric perforation

Selected References

1. Kahrilas PJ, Quigley EMM. Clinical esophageal pH recording: A technical review for practice guideline development. Gastroenterology 1996; 110:1982-1996.
   This review presents the official recommendations for the use of esophageal pH monitoring by the American Gastroenterological Association.
2. Johnson LF, DeMeester TR. Development of the 24-hour intraesophageal pH monitoring composite scoring system. J Clin Gastroenterol 1986; 8(suppl 1):52-58.
   In this article Johnson and DeMeester present the logic behind the derivation of their classic acid-reflux composite scoring system.
3. Bremner RM, DeMeester TR, Stein HJ. Ambulatory 24 hour esophageal pH monitoring—What is abnormal? In: Richter JE, ed. Ambulatory Esophageal pH Monitoring: Practical Approach and Clinical Applications. 2nd ed. Baltimore: Williams & Wilkins, 1997:77-95.
   The chapter provides a thorough explanation about the derivation of "normal" values for pH monitoring. Receiver operating curves are provided to demonstrate how appropriate cut-off values were chosen. The textbook in general is an excellent resource for anyone interested in learning more about ambulatory pH monitoring.
4. Galmiche JP, Scarpignato C. Esophageal pH monitoring. In: Scarpignato C, Galmiche JP, eds. Functional evaluation in esophageal disease. Front Gastrointest Res 1994; 22:71-108.
   This is a detailed review that discusses aspects of pH monitoring such as diet and activities during the monitoring period, reproducibility of diagnosis, and an overview of the various reference values.
5. Ergun GA, Kahrilas PJ. Clinical applications of esophageal manometry and pH monitoring. Am J Gastroenterol 1996; 91:1077-1089.
   This is a review focusing on the clinical applications of both esophageal manometry and pH monitoring. Attention is given to the role of pH monitoring in evaluation and management of gastroesophageal reflux and its complications.

6. Waring JP, Hunter JG, Oddsdottir M, Wo J, Katz E. The preoperative evaluation of patients considered for laparascopic antireflux surgery. Am J Gastroenterol 1995; 90:3538.
In this prospective study 88 patients underwent an esophagogastroduodenoscopy (EGD), esophageal manometry, and a 24-hour ambulatory esophageal pH monitoring as part of a preoperative evaluation for antireflux surgery. The role of each procedure in affecting surgical decision making was then reviewed.

CHAPTER 9

Gastrointestinal Foreign Bodies

Patrick G. Quinn

Epidemiology

Major Risk Groups

- Children ages 15 years with accidental ingestion (age range 3 months to 12 years)
- Adults with food impaction (risk factors: dentures, prior esophageal symptoms/disease)
- Adults with true foreign bodies (risk factors: mental retardation, psychiatric care, incarcerated)

Presentation

- Sudden onset odynyphagia or dysphagia with sialorrhea (warning: symptoms may be minimal or nil)
- Maintain a high index of suspicion

History

- Exact timing of ingestion
- Description of ingested objects, interviews with witnesses
- Prior esophageal symptoms or disease

Physical Exam

- General state
- Respiratory status
- Difficulty with secretions or swallowing
- Dentition
- Palpable throat, neck, or abdominal masses
- Crepitus in soft tissue of neck
- Peritonitis or signs of small bowel obstruction

Initial Radiographic Evaluation

- PA and lateral neck films if neck pain or suspicion of bone ingestion
- Chest x-ray (even in children since foreign body may have minimal symptoms)
- Not needed if meat impaction in clinically stable adult
- Additional films to survey mouth to anus in children to exclude multiple objects
- Always consider possibility of a second or additional foreign body
  - High false negative rates, so absence of findings not reliable

Gastrointestinal Endoscopy, edited by Jacques Van Dam and Richard C. K. Wong.

Table 9.1. Types of foreign bodies

Fish bones	Tacks	Toothbrush
Keys	Buttons	Taco shells
Watches	Pits	Toothpicks
Nuts	Metal clips	Drug filled packets
Seeds	Pen caps	Dental bridges
Marbles	Pins	Crayons
Antennae	Dentures	Utensils
Batteries:	Coins:	
Button	Dime, 17 mm	
AA	Penny, 18 mm	
AAA	Quarter, 23 mm	
Transistor		

- Many commonly ingested objects are radiolucent metal bottle tops, wood, glass, and plastic
- Migration may occur in interval between film and endoscopy
- Avoid barium in high grade esophageal obstruction due to risk of aspiration
- Small amounts of dilute barium or gastrograffin can be used to locate radiolucent objects (use of gastrograffin may facilitate endoscopy shortly after x-ray done)

Indications for Removal

Emergent

- Respiratory distress
- Inability to swallow secretions
- Perforation
- Button battery in esophagus

Urgent

- Coin in the proximal 2/3 of esophagus
- Sharp objects in esophagus or stomach such as fish or chicken bones, razor blades, glass shards, toothpick

Elective

- Any foreign body associated with symptoms
- Coin in distal esophagus >24 hours
- Esophageal meat impaction >12 hours
- Button batteries in stomach >48 hours
- Object larger than 2.5 cm
- Object longer than 6 cm in children or 10-13 cm in adults
- Gastric retention of small object longer than 23 days
- Coin in stomach > 4 weeks
- Button battery in small bowel or colon > 10 days

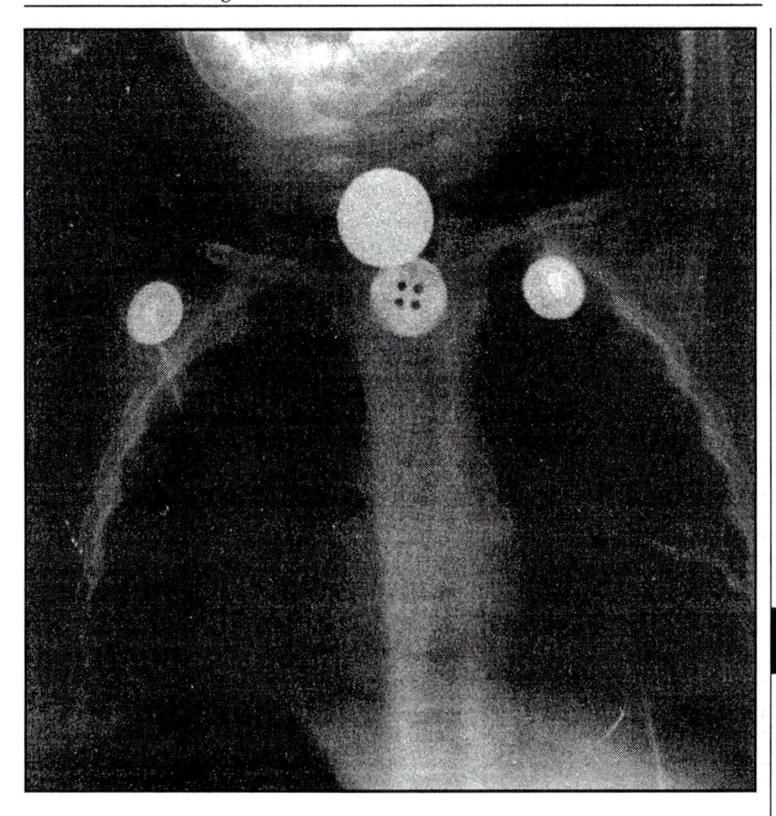

Figure 9.1. AP chest radiograph of a 3 month old child who had ingested both a button and then a penny which resulted in difficulty swallowing secretions. Note that objects in the esophagus are coronal in the AP projection.

Options for Removal, Nonendoscopic

Esophagus

- Promote passage
    - Trial of anxiolytics, analgesics, spasmolytics (diazepam, glucagon, scopolamine, nitroglycerin, nifedipine)
    - Consider trial in first 12 hours with intravenous hydration
    - Gas forming agents (EZ gas or cola)
    - Avoid if rigid or fixed obstruction or object in the proximal 1/3 of esophagus
        - Risks: there has been a report of esophageal perforation
    - Dissolution of meat with papain
        - Risks: contact with inflamed mucosa caused mucosal damage in animal studies, and esophageal perforation has been reported

Table 9.2. Iatrogenic foreign bodies

- Gastrostomy tubes
- Biliary stents
- Sucralfate
- Dental drills
- Esophageal stents
- Impacted Dormia baskets
- Surgical gauze
- Cyanoacrylate concretions

Table 9.3. Bezoars

Phytobezoar:	Vegetable material
Diospyrobezoar:	Persimmon berries
Trichobezoar:	Hair
Lactobezoar:	Milk

- Blind passage of bougie (Hurst blunt tip) dilator
    - Risk: esophageal perforation, but recent report of emergency room use for pediatric patients suggests minimal risk
- Promote expulsion
    - Use of **emetics**
        - Risks: uncontrolled, could cause rupture, perforation, or aspiration
    - **Extraction with foley balloon catheter** under fluoroscopic guidance, inflate with contrast, large case series with high success rate (92% of 229 cases) Criteria: blunt objects in place < 12 hours. Risk: aspiration or tracheal obstruction if object falls into hyopharynx, and have McGill forceps and laryngeoscope available. Some argue only use if endoscopy not available.

9

Options for Removal, Endoscopic

General

- Cooperative or well sedated patient required to maximize safety. In children general anesthesia and endotracheal intubation often needed
- Always visualize path of endoscope. It may be risky to attempt blind intubation of esophagus in this setting if hypopharyngeal object present
- Consider practice with an object similar to the one ingested to facilitate the procedure
- If repeated intubation required, consider overtube use
- Once object removed, esophagus may be dilated if stricture present and minimal inflammation
- Narcotic-filled condoms or wrapped in plastic should not be removed endoscopically due to risk of rupture

Table 9.4. Endoscopic accessories for foreign body removal

Rattooth forcep	Two channel therapeutic scope
Alligator forcep	Banding overtube
Tripod	Enteroscope (gastric) overtube
Snares	Endoscopic suture scissors
Roth retrieval net	Latex hood or bell
Bipolar snare	Band ligator
TTS balloons	Dormia basket

Table 9.5. Sedation guidelines

General anesthesia:
- Infants
- Children
- Deliberate ingestors
- Difficult objects

If ET tube not used: Trendelenberg position
ET tube required for button battery removal using TTS balloon

Esophagus

- **Coin**
  - Grasp with rattooth or alligator forceps
  - Orient in coronal plane and withdraw scope
- **Food bolus**
  - If scope can passed alongside bolus and distal esophagus not obstructed, gentle pressure may pass object into stomach (always push from right side at GE junction)
  - If recently ingested and firm, grasp with snare or Dormia basket and remove in toto
  - Band ligator tip may allow suction to remove object in one piece especially if object fragments when grasped
- **Button battery**
  - Usually shows a double density or halo on X ray
  - Dormia basket, tripod, or Roth retrieval net can be used to grasp object
  - Inflating a TTS balloon below object and withdrawal along with scope possible, but does not allow control of object in hypopharynx, so may pose an aspiration risk
  - Following removal of button battery get Barium swallow at 2436 hours to rule out fistula and at 1014 days to rule out stricture
  - If tissue damage present, consider antiobiotic treatment
- **Round objects**
  - Dormia basket, tripod, or Roth retrieval net can be used to grasp object
- **Sharp objects**
  - If point oriented cephalad, grasp object, pass into stomach, reorient, then remove with sharp end trailing

Table 9.6. Anatomical considerations

Esophagus:	
	Zenker's diverticulum
	Aortic arch
	Left mainstem bronchus
	Gastroesophageal junction
Gastric:	
	Pylorus
Small bowel:	
	Duodenal sweep
	Meckel's diverticulum
	Ileocecal valve
Colon:	
	Appendix
	Anus

Gastric

- Techniques similar to esophagus
  - For sharp objects consider use of overtube or latex bell if sharp sides to limit laceration risk at GE junction and upper esophageal sphincter
  - Clear band ligator tip on scope may facilitate mucosal inspection for glass or clear objects
  - **Atropine** may decrease gastric secretions
  - **Glucagon** may decrease lower esophageal sphincter pressure and decrease gastric contractions
  - **Bezoars** (see Table 9.3)
    - Usually diagnosed on upper GI series
    - Small bowel obstruction most common reported complication
    - Enzymatic digestion described with papain, acetylcysteine, cellulase, or pineapple juice
    - Clear liquid diet and promotility agents may be used prior to endoscopy and should help between repeated endoscopy sessions
    - Lavage with saline, 0.1N HCl or sodium bicarbonate reported
    - Disrupt with rattooth, snare, tripod, Dormia basket
    - Consider use of overtube 2 channel therapeutic endoscope or band ligator tip to remove fragments; even Nd:YAG laser use reported

Colon

- Diagnosis often obscure since only systemic symptoms or vague abdominal pain
- Most patients do not recall foreign body ingestion (e.g., toothpicks)
- Generally pointed objects will pass without symptoms if entrained in stool
- Possible complications: perforation, peritonitis, vascular compromise
- Indications for colonoscopic extraction: failure to traverse ileocecal valve, obstruction, contained perforation, presence of pointed or elongated object
- Consider: surgical backup, liberal antibiotics, follow-up barium enema or CT
- Use rattooth forceps, Dormia basket or snare
- If in rectum consider anal or perineal block and rigid scope to protect sphincter

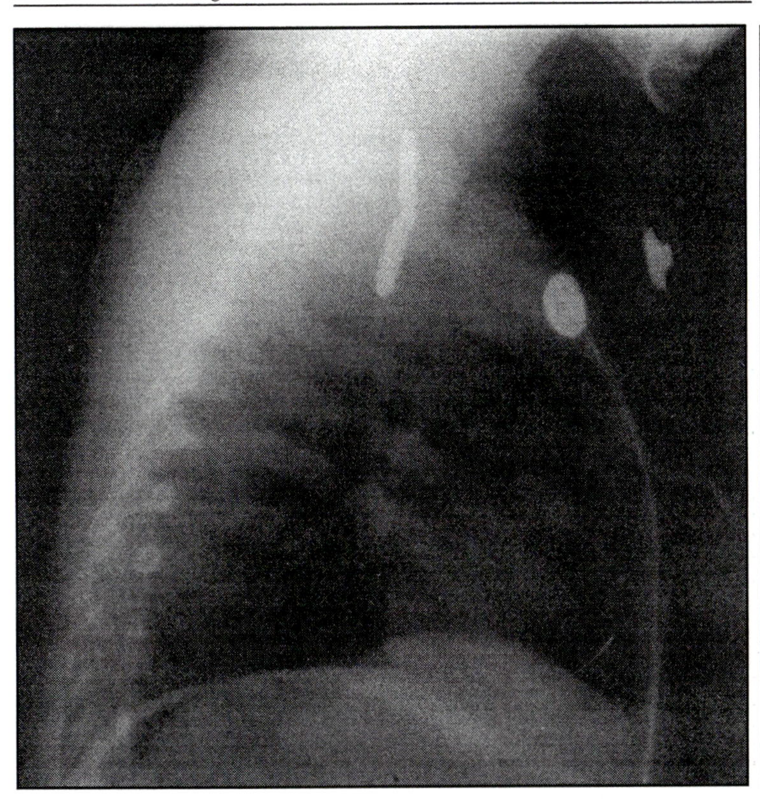

Figure 9.2. Lateral view of same child shown in Figure 9.1. Note that esophageal objects appear on end in the lateral projection. Tracheal objects would appear coronal in the lateral projection.

Options if Unable to Remove Foreign Body at Endoscopy

- 80-90% of gastric foreign bodies will pass spontaneously within four to six days
- Object in esophagus less than 2 cm diameter and 5 cm long can be pushed into stomach
- High roughage diet; screen stools for one week
- If not passed, recheck abdominal radiograph weekly
- For sharp objects or narcotic-filled condoms follow daily radiographs
- Repeat removal attempt for:
  - Coin or blunt object stationary in stomach for 34 weeks
  - Sharp object stationary in small bowel more than 35 days
- Consider surgical removal for symptoms if fever, vomiting, or abdominal pain, or if signs of gastrointestinal bleeding, or if object not progressing and not endoscopically accessible

Selected References

1. Guideline for the management of ingested foreign bodies. Gastrointest Endosc 1995; 42:622625.
2. Webb WA. Management of foreign bodies of the upper gastrointestinal tract: update. Gastrointest Endosc 1995; 41:3951.
3. Quinn PG, Connors PJ. The role of upper endoscopy in foreign body removal. In Blades EW, Chak A, eds. Gastrointest Clin NA 1994; 4:571593.
4. Ginsberg GG. Management of ingested foreign objects and food bolus impactions. Gastrointest Endosc 1995; 41:3338.
5. Faigel DO, Stotland BR, Kochman ML et al. Device choice and experience level in endoscopic foreign object retrieval: An in vivo study. Gatrointest Endosc 1997; 45:490492
6. Emslander HC, Bonadio W, Klatzo M. Efficacy of esophageal bougienage by emergency physicians in pediatric coin ingestion. Ann Emerg Med 1996; 27:726729.
7. Andrus CH, Ponsky JL. Bezoars: Classification, pathophysiology, and treatment. Am J Gastroenterol 1988; 83:476478.

CHAPTER 10

Endoscopic Therapy for Nonvariceal Acute Upper GI Bleeding

Peder J. Pedersen and David J. Bjorkman

Introduction

Upper gastrointestinal bleeding (UGIB) remains a significant medical problem. Recent years have brought many exciting advances in diagnostic and therapeutic options for UGIB. It should be noted that 80-85% of patients with UGIB experience spontaneous resolution of their hemorrhage without therapeutic intervention.

UGIB accounts for 250,000-300,000 hospital admissions per year; $2.5 billion are spent annually on patients with UGIB.

Mortality remains 5-10% despite advances in care. Deaths are now seen in older patients with more comorbidity.

Anatomy

UGIB occurs mostly in the stomach and proximal duodenum (bulb and second portion). Peptic ulcer disease is, by far, the most common cause of acute UGIB (Table 10.1). Incidence of variceal bleeding varies with the socioeconomic demographics of the patient population.

Other Sites

- Esophagus
- Third portion of the duodenum
- Bile duct (hemobilia)
- Pancreas (hemosuccus pancreaticus)

Evaluation of Patient with UGIB

Initial Management (Occurs Simultaneously)

- Hemodynamics
  - Shock (SBP less than 90) suggests about 50% blood volume loss
  - Orthostatic hypotension (SBP falls to less than 90 with standing) suggests 25-50% loss
  - Orthostatic fall in systolic blood pressure of greater than ten or rise in pulse of greater than twenty suggests 20-25% loss

Note that these are general guidelines and various medical conditions such as advanced age, or medications such as beta blockers can affect the above changes.

- **IV access** – 2 large bore IVs of 18 gauge or larger
- Focused history and physical
  - Orthostatic symptoms
  - Hematemesis/melena/hematochezia

Gastrointestinal Endoscopy, edited by Jacques Van Dam and Richard C. K. Wong.

Table 10.1. Sources of acute upper GI bleeding

Source	%
Peptic ulcer disease	40-50%
Duodenal ulcer	24-27%
Gastric ulcer	13-21%
Mallory Weiss tear	7-13%
Gastritis	3-23%
Duodenitis	0-5%
Esophagitis	1-2%
Vascular malformations	0.5-2%
Malignancy	1-3%
Other	1-6%
No lesion identified	0-7%
Varices (esophageal or gastric)	10-31%

- Previous illnesses especially cardiac, hepatic, and renal disease; malignancy; substance abuse.
- Medications especially nonsteroidal antiinflammatory drugs (NSAIDs) including over-the-counter, combination remedies and daily aspirin; steroids; and anticoagulants such as warfarin and heparin

- **Volume resuscitation** with crystalloid such as normal saline or lactated Ringers. Goal is resolution of hemodynamic instability
- Laboratories: CBC, platelet count, PT, type and cross. Realize that acutely red blood cells and plasma are lost concurrently so hematocrit/hemoglobin remain constant in spite of significant blood loss

Initial Localization (Clinical)

- **Melena** (black, unformed, tarry, malodorous stool) 50-100 cc of blood loss that usually occurs above the ligament of Treitz.
- **Hematochezia**. About 10% of patients with bright red blood per rectum have UGIB. In this situation, the volume is large (1000cc or more blood loss) and bleeding is generally accompanied by hemodynamic instability.
- **Hematemesis/Coffee Grounds**
  - May or may not indicate active bleeding
  - "Coffee grounds" are formed when blood interacts with acid.

Gastric Lavage: Helpful or Necessary?

- 80% sensitive for active UGI bleeding
- The presence of bile may increase negative predictive value for active UGI bleeding (proximal to ligament of Treitz)
- Helps determine if bleeding is active
- Clearance of blood and clots from stomach (requires a large bore 34 F Ewald tube)

Are chemical tests for blood, such as guaiac or gastrocult helpful? In general chemical tests for blood in the stool (**guaiac**) or gastric aspirate (**gastrocult**) are not helpful. Both have high false positive and false negative rates. False positive gastrocult can occur with trauma due to placement of the NG tube. False negative results may be obtained if bleeding is distal to the pylorus or the bleeding has stopped. The real question to be answered is whether active bleeding has occurred or is still occurring. This is a clinical question answered by the presence of coffee grounds or gross blood seen in the gastric aspirate or the vomitus, and hemodynamic changes with melena or hematochezia.

Risk stratification/prognosis based on clinical grounds. Worse prognosis in:

- Age greater than 60
- Major comorbidity
- Already hospitalized for nonGI bleeding illness
- Shock on presentation
- Rebleeding or severe persistent bleeding

Indications/Contraindications for Endoscopy in UGIB

Adequate resuscitation should always proceed endoscopy as upper endoscopy and sedation can cause circulatory collapse if the patient is not adequately resuscitated.

Indications for Urgent/Emergent Endoscopy

- Severe or ongoing bleeding
- Recurrent bleeding
- Bleeding in a patient hospitalized for other illness
- Severe comorbid illness (This decreases the ability of the patient to tolerate recurrent bleeding.)
- Possible varices
- In general, high clinical risk

Possible indications for elective endoscopy (i.e., admit and perform endoscopy at the earliest convenience)

- Low clinical risk
- No severe comorbidities
- Low volume bleed
- Hemodynamically stable or easily stabilized

Early (pre-admission) endoscopy may:

- Identify bleeding lesions
- Predict risk of rebleeding (see below) by endoscopic appearance of lesions.
- Allow early discharge or outpatient care of selected patients
- Reduce costs of care

Contraindications to Endoscopy

- Patient refusal
- Uncorrected coagulopathy (relative contraindication)
- Unstable cardiac disease or recent MI (relative contraindication)

Endoscopy Equipment

Therapeutic Gastroscope

- Two working channels of a single extra large channel. In either case, a channel must be large enough to accommodate a 10 F catheter (usually 3.7 mm)
- Allows suction of some clots or blood. (Although, if massive amounts of retained blood and clots are present, they can not be cleared through the endoscope and vigorous gastric lavage is indicated)
- Allows passage of large-sized hemostasis equipment

Hemostasis Equipment

- Injection catheters: allow an injection of epinephrine or other solutions such as sclerosants or desiccants
- Heater probe
- Bicap
- Laser fibers
- Argon plasma coagulator
- Hemoclips
- Band ligator(s)

Endoscopic Findings

Peptic Ulcer Disease-Most Common Finding

- Stigmata of recent hemorrhage are associated with varying risks of recurrent bleeding and mortality.
  - Clean-based ulcer with whitish or yellowish base and no other findings
  - Flat pigmented spot: a flat red, purple or black spot in the ulcer base, without clot or elevation
  - Adherent clot: blood clot firmly affixed to the ulcer crater that stays in place despite vigorous flushing.

Note: there is controversy concerning whether forcibly removing an adherent clot is clinically indicated or safe. (See below, Endoscopic Therapy and Management of UGIB)

  - Nonbleeding visible vessel: translucent to white, red or purple raised lesion in the ulcer crater. This actually represents a fibrin plug or clot in an arterial wall defect.
  - Active bleeding

Rebleeding Rates

Other Lesions

- Stress gastritis: occurs in severely ill or injured patients.
- Vascular ectasia
- Dieulafoy's lesion
- Esophagitis/esophageal ulcer
- Tumors
- Mallory-Weiss tear
- Stomal (a.k.a. anastomotic or marginal) ulcers
- Chronic gastritis (*Helicobacter pylori*) is an uncommon cause of significant bleeding in the absence of an ulcer.
- Chemical gastropathy (NSAIDs, bile) may cause chronic low-grade blood loss.

Endoscopic Therapy and Management of UGIB

Endoscopic Therapy

- Controls bleeding in greater than 90% of patients
- Reduces rebleeding
- Reduces morbidity
- Improves mortality

Table 10.2. Rebleeding and mortality rates for ulcers with endoscopic stigmata

	Active Bleeding	Visible Vessel/Bleeding Clot
Adherent clot	40-90%	35-60%
Flat spot	15-35%	5-15%
Clean base	0-5%	5-20%
	5-20%	0-10%
	0-7%	0-2%

Appropriate lesions for therapy include the major stigmata of recent hemorrhage including adherent clot, nonbleeding visible vessels, and active bleeding.

All methods have similar efficacy.

Technique

- **Injection therapy**. This may work due to spasm of the artery or dehydration of the vessel or simply secondary tamponade.
  - Epinephrine at a dilution of 1/10,000
  - Normal saline
  - Polidocanol
  - Absolute alcohol
- **Thermal therapy**. This welds the vessel walls together. To do so requires a combination of pressure to coapt the vessel walls and heat to weld them together.
  - Heater probe–direct heat application
  - Multipolar coagulation–uses electrical resistance of tissues to cause thermal injury
- **Laser and argon plasma coagulator**. Use heat to dehydrate the tissues and cause coagulation. Now rarely used because of expense and logistical issues.
  - **Management of adherent clot**: controversy exists as to whether a clot should be forcibly removed.
  - Arguments for removal:
    - Allows the endoscopist to visualize the ulcer base and the specific bleeding lesion
    - It is not thought to increase the risk of complications
    - Can allow more specific, directed therapy
    - There is, for example, a visible vessel (high-risk lesion) or a flat spot (low-risk lesion) below the clot.
- Arguments against removal include that you are removing a naturally occurring hemostatic plug and may precipitate hemorrhage
- If performed, we recommend that the endoscopist inject the ulcer base with epinephrine first, then remove the clot, and be prepared to immediately apply thermal therapy to any high-risk lesion

General Management Strategy

In general, we use a combination of injection therapy followed by thermal therapy, usually using a multipolar probe. This strategy, however, is often times modified by the clinical situation or the accessibility of the lesion.

Table 10.3. Endoscopic therapy is beneficial in acute UGIB

Effectiveness of thermal therapy (heater probe, BICAP) for active UGIB or non-bleeding visible vessel

	Control	Thermal Therapy
Initial hemostasis (%)	13-20%	85-100%
Re-bleeding (%)	37-72%	15-40%

Effectiveness of injection therapy (epinephrine, sclerosants, absolute alcohol, saline) for UGIB

	Control	Injection Therapy
Initial hemostasis (%)	0%	62-100%
Re-bleeding	43%	6-25%

Medical Therapy (Acid Suppression)

The efficacy of **acid suppression** in treating UGI bleed remains controversial. In general, we feel that it will probably decrease rebleeding and be additive to endoscopic therapy if employed with high risk lesions (adherent clot, visible vessel, and active bleeding) and used in adequate doses (omeprazole 40-60 mg PO bid or the equivalent dose of other proton pump inhibitors).

Complications

In general, complications of endoscopic management of UGIB are similar, if slightly more frequent than elective diagnostic or therapeutic upper endoscopy. Ways to decrease chances of complications include:

- Adequate resuscitation
- Consider airway control (intubation) if you suspect potential aspiration due to decreased mental status, massive bleeding or hematemesis.

10

Selected References

1. Consensus Development Panel. Therapeutic endoscopy and bleeding ulcers. JAMA 1989; 262:1369-1372.
   A consensus statement on the role of endoscopy in acute UGIB.
2. Fleischer D. Endoscopic therapy of upper gastrointestinal bleeding in humans. Gastroenterology 1985; 90:217-234.
   A review of various endoscopic modalities available to treat acute UGIB.
3. Hay JA, Lyubashevski E, Elashoff J et al. Upper gastrointestinal hemorrhage clinical guideline—Determing the optimal hospital length of stay. Am J Med 1996; 100:313-322.
   The authors developed a risk stratification scoring system and evaluate its impact on length of stay for acute UGIB.
4. Kovacs TOG, Jensen DM. Therapeutic endoscopy for non-variceal upper GI bleeding. In: Tiller MB, Gollan JL, Steer ML et al, eds. Gastrointestinal Emergencies, 2nd ed. Baltimore: Williams and Wilkins, 1997:181-198.
   This chapter systematically reviews all of the therapeutic options for endoscopic treatment of non-variceal GI bleeding and compares them to each other, including outcomes and rates of re-bleeding. The authors also give recommendations on therapeutic approaches.

5. Laine L. Multipolar electrocoagulation in the treatment of peptic ulcers tith nonbleeding visible vessels: a prospective controlled trial. Ann Intern Med 1989; 110:510-524.
A randomized controlled trial that showed about a 50% relative risk reduction in rebleeding (41% in controls vs. 18% in treated patients), as well as decreased need for surgery and shorter hospital stays for patients treated with multipolar electrocoagulation.
6. Laine L. Multipolar electrocoagulation versus injection therapy in the treatment of bleeding peptic ulcers. Gastroenterology 1990; 99:1303-1306.
A randomized comparative trial that evaluated multipolar electrocoagulation and injection of absolute alcohol. The study showed similar rates of initial hemostasis and all other criteria, including rebleeding rates and mortality.
7. Laine L. Upper gastrointestinal tract hemorrhage. West J Med 1991. Sep; 155:274-279.
A classic review on management of patients with acute UGIB.
8. Laine L, Peterson WL. Bleeding peptic ulcer. N Engl J Med 1994; 331:717-727.
This article reviews the clinical risk of peptic ulcer disease and includes a review of various studies that have evaluated stigmata recently.
9. Lichtenstein DR, Berman MD, Wolfe MM. Approach to the patient with acute upper gastrointestinal hemorrhage. In: Tiller MB, Gollan JL, Steer ML et al, eds. Gastrointestinal Emergencies, 2nd ed. Baltimore: Williams and Wilkins, 1997:99-129.
This extensive review of the initial evaluation and management of patients with acute upper GI bleeding discusses clinical risk stratification, prognosis, and initial management approach to patients with acute GI bleeding.
10. Pedersen PJ, Bjorkman DJ. Omeprazole for bleeding PUD: Do we finally have evidence for effective medical therapy? Am J Gastro 1998; 93 (9):1583-1585.
This article reviews recent studies on acid suppression as medical therapy for bleeding peptic ulcer disease.
11. Rockall TA, Logan FRA, Devlin HB et al. Risk assessment after acute upper gastrointestinal haemorrhage. Gut 1996; 38:316-321.
This study developed a clinical risk stratification scheme for patient triage in acute nonvariceal upper GI bleeding.
12. Rockall TA, Logan RFA, Devlin HB et al. National audit of acute upper gastrointestinal haemorrhage. Selection of patients for early discharge or outpatient care after acute upper gastrointestinal haemorrhage. Lancet 1996; 347:1138-1140.
A report of retrospective application of a clinical risk stratification scoring scheme in acute UGIB.
13. Silverstein FE, Gilbert DA, Tedesco FJ et al. The national ASGE survey on upper gastrointestinal bleeding. II. Clinical prognostic factors. Gastrointest Endosc 1981; 27:80-93.
Part II in a 3-part series, this paper presents data on clinical risks stratification in patients with acute UGIB.
14. Standard of Practice Committee. The role of endoscopy in the management of non-variceal acute upper intestinal bleeding. Gastrointest Endosc 1992; 38:760-764.
This is an American Society for Gastrointestinal Endoscopy position paper on endoscopic management of acute upper GI bleeding.
15. Wara P. Endoscopic prediction of major re-bleeding: A prospective study of stigmata of hemorrhage in bleeding ulcer. Gastroenterol 1985; 88:1209-1214.
This is an endoscopic study of stigmata of recent hemorrhage looking at rates of rebleeding.

CHAPTER 11

Endoscopic Management of Lower GI Bleeding

Sammy Saab and Rome Jutabha

Introduction

Colonoscopy is the procedure of choice for the diagnosis and treatment of lower gastrointestinal bleeding (LGIB) with the cecum being reached in over 95% of patients. It permits direct visualization of the mucosa, provides the opportunity for local therapy and allows for the procurement of pathologic specimens. A rational differential diagnosis for the etiology of LGIB can be generated from a careful history and focused physical examination (Table 11.1). For example, a young patient with hematochezia (bright red blood per rectum) is more likely to have inflammatory bowel disease, hemorrhoids, or colonic infection. In contrast, patients over 50 years of age are more predisposed to colonic diverticulosis, angiodysplasia, polyps or carcinoma. Upper endoscopy should be considered in any patient with severe hematochezia and signs of hemodynamically significant hemorrhage in order to exclude a briskly bleeding upper GI source.

Indications for Colonoscopy During Acute Lower Gastrointestinal Hemorrhage

- Identify cause and location of lower gastrointestinal hemorrhage
- Perform endoscopic therapy, hemostasis
- Direct further medical, surgical or radiographic management
- Obtain mucosal biopsy for histopathologic examination or culture

Contraindications

- Medically unstable patient
- Active myocardial ischemia/infarction
- Severe thrombocytopenia (platelet count < 25,000)
- Severe coagulopathy (prolonged prothrombin time, e.g., INR > 1.5)
- Inability to visualize the lumen/poor patient preparation
- Fulminant colitis, toxic megacolon or bowel perforation

Patient Preparation

- Prior to emergent colonoscopy, all patients must be hemodynamically stable or attempts at resuscitation must be performed.
- Insert two large bore intravenous catheters
- Obtain complete blood count, coagulation studies and serum chemistries. Type and crossmatch. Correct abnormalities such as anemia, thromobocytopenia, and coagulopathy
- If practical, prepare patient using oral purge

Gastrointestinal Endoscopy, edited by Jacques Van Dam and Richard C. K. Wong.

Table 11.1. Common causes of lower gastrointestinal bleeding

-Diverticulosis
-Angiodysplasia
-Radiationinduced telangiectasia
-Infectious
-Ischemic
-Idiopathic; inflammatory bowel disease
-Neoplastic:
    Polyp
    Carcinoma
-Other:
    Hemorrhoid
    Ulcer
    Postpolypectomy

Relevant Anatomy

- **Anatomy** The colon is divided anatomically into four areas: cecum/ascending colon, transverse colon, descending colon, and sigmoid colon. The total length of the colon is generally between 100 and 150 cm.
- **Vascular supply** The colonic blood supply can be quite variable. Nevertheless, some generalizations are possible: The superior mesenteric artery is located at the level of the first lumbar vertebrae and provides branches that supply the ascending colon and the transverse colon. The inferior mesenteric artery is located at the level of fourth lumbar vertebrae and supplies the descending and sigmoid colon. The splenic flexure is supplied by distal branches from the both the superior and inferior mesenteric arteries and thus is particularly susceptible to ischemic injury (watershed area).
- Other **diverticula** are mucosal herniations through the muscular wall of the colon and measure between 0.5 to 1 cm in length. They are located predominantly in the sigmoid colon and tend to occur at weakened areas where arterial branches penetrate the submucosa. Angiodysplasia refers to dilated tortuous submucosal veins whose walls are composed of endothelial cells lacking in smooth muscle.

Equipment

- Endoscopes differ in their length and diameter but incorporate a biopsy channel which allows for water irrigation and the passage of biopsy forceps and hemostasis accessories (see below).
- A vacuum pump for suctioning blood, debris and lavage fluid
- An air/water pump for lens cleansing and the infusion of air into the colon.

Accessories for Hemostasis

- Epinephrine injection
    - Local injection through a needle catheter (sclerotherapy needle) can provide hemostasis in many situations (e.g., ulcer, diverticulum, angiodysplasia and postpolypectomy hemorrhage). A typical dose of epinephrine is 5-10 ml of dilute (1:10,000) solution.
- Bipolar electrocoagulation

- Two **bipolar electrocoagulation** devices are available: Gold ProbeÆ (Microvasive Inc.) and BICAPÆ probe (ACMI Corp.). The Gold Probe has been successfully used in the local treatment of bleeding diverticulum. With a setting of 34 on a 50 watt bipolar electrocoagulation generator, 6 to 18 one-second pulses are applied to the visible vessel at the neck of a diverticulum. Results are best with the large, 3.2 mm probe. The BICAP probe has been studied in the treatment of bleeding hemorrhoids and radiation proctopathy. Using a power setting of 5 with pulse duration of 2 sec, successful hemostasis can be achieved.

- **Thermocoagulation**
  - Heater probes are available in two diameters: 2.4 mm and 3.2 mm. The larger size probe is more effective at achieving hemostasis. Settings used are between 10 and 15 joules. The heater probe is used for the local treatment of angiodysplasia, radiation telangiectasia and ulcers.
- **Nd:YAG laser**
  - The laser is effective in treating radiation proctopathy and angiodysplasia. However, its applicability is limited by its large size (nonportable), special water and electrical connections, need for special training to operate, requirement for eye protection, and high cost.
- **Band ligation**
  - Typically used in the treatment of internal hemorrhoids but has been used successfully for treating postpolypectomy bleeding and diverticular bleeding. The technique involves a plastic device with rubber bands positioned over the end of the endoscope. A trigger device placed near the hand piece sequentially deploys the bands around the designated target.

Technique

- Preparation
  - Successful identification and hemostasis of a bleeding source requires adequate visualization of the colonic mucosa. One method to achieve bowel cleansing is at least 4 L of a polyethylene glycolbased oral purgative given over 2 hours. Ten milligrams of metoclopramide can be administered at the start of the purge to facilitate intestinal transit and minimize the risk of nausea and vomiting. For noncompliant patients, a nasogastric tube can be used to help administer the solution.
- Position
  - Patients should be placed in the left lateral decubitus position. A rectal examination is performed prior to the insertion of the endoscope to lubricate the anal canal and evaluate for rectal masses or retained stool.
- Sedation
  - Analgesic (narcotics) and anxylotic (benzodiazepines) agents are administered intravenously to patients who are hemodynamically stable in increments every 23 min as needed. Agents typically used in conscious sedation include midazolam (diazepam may be used), meperidine, or fentanyl.

Outcome

- Extent. The cecum can be reached in up to 95% of patients presenting with significant hematochezia undergoing colonoscopy.
- Yield. A diagnosis can be made in up to 80% of patients. In patients older than 50 years of age, the most common causes are angiodysplasia and diverticulosis.

In younger patients, colitis and hemorrhoids are common causes.

- Hemostasis
  - **Hematochezia** tends to be self-limited and resolves in most patients without therapeutic intervention. Endoscopy allows for diagnosis and local therapy when necessary.
  - The neck of bleeding diverticula can be injected with epinephrine or coagulated with a bipolar device such as a Gold Probe (Microvasive Inc.). Using large probes, no rebleeding was found in one series nine months after therapy.
  - Angiodysplasia can be successfully treated with endoscopic coagulation (with bipolar probe or heater probes), injection sclerotherapy, and argon laser coagulation. However, recurrent bleeding after local therapy can occur in up to 30% of patients.
  - Neoplasms are generally not treated endoscopically. Because of friability and size, there is significant risk of inducing more bleeding or causing a perforation with endoscopic therapy. Colonoscopy is useful for obtaining tissue for histopathologic examination (diagnostic biopsies).
  - Hemorrhoids can be successfully treated endoscopically with heater probe, electrocoagulation and rubber band ligation.
  - Radiation proctitis is amendable to treatment with bipolar electrocoagulation, heater probe coagulation and Nd:YAG laser.

Complications

- Hypotension. Hypotension can be related to the hypovolemic status of a patient and/or the use of agents used to promote conscious sedation. Treatment includes volume resuscitation with intravenous (IV) fluid and/or packed red blood cells and IV naloxone (0.4 mg), an opioid antagonist.
- Respiratory compromise. When it occurs, respiratory compromise is most likely related to the analgesics and anxiolytics used in conscious sedation. Narcan and flumazenil are agents used to reverse the effects of narcotics and benzodiazepines, respectively. Flumazenil is administered in 0.2 mg increments intravenously over 15 sec. It can be repeated every minute up to 1 mg. Supplemental oxygen should be provided.
- Allergic reaction. An allergic reaction can be caused by any of the sedative medications and is manifested by the onset of a rash or itching near the IV catheter administration site. Avoid further sedation and administer IV antiH1 histamine blockers (e.g., diphenhydramine). Consider alternative sedative.
- Discomfort. Despite the use of conscious sedation, patients may complain of abdominal discomfort due to colonic distension or stretching of the bowel (secondary to the colonoscope, or air insufflation). Mild to moderate discomfort is not unusual; however, severe pain signifies extreme distension or stretching with a high risk of perforation. Severe pain warrants cessation of the maneuver causing the pain or the procedure itself.
- Perforation. Perforation may be caused by the tip of the endoscope or it may occur at a fixed area (e.g., hepatic or splenic flexure, or in an area of adhesions) when advancing the colonoscope despite the formation of a loop. The likelihood of perforation is higher when the colonic wall is weakened by inflammation or tumor. Delayed perforation may also occur 24 days following polypectomy because of transmural tissue injury at the polyp resection site or contralateral wall (if in contact with the polyp during electrocautery).

Selected References

1. Zuccaro G Jr. Management of the adult patient with acute lower gastrointestinal bleeding. Am J Gastroenterol 1998; 93:1202-1210.
2. Waye JD, Bashkoff E. Total colonoscopy: Is it always possible? Gastrointest Endosc 1991; 37:152-154.
3. Carbatzas C, Spencer GM, Thorpe SM et al. Nd:YAG laser treatment for bleeding from radiation proctitis. Endoscopy 1996; 28:497-500.
4. Naveau S, Aubert A, Poynard T et al. Longterm results of treatment of vascular malformations of the gastrointestinal tract by neodymium YAG laser photocoagulation. Dig Dis Sci 1990; 35:821-826.
5. Jensen DM, Machicado GA, Cheng S et al. A randomized prospective study of endoscopic bipolar electrocoagulation and heater probe treatment of chronic rectal bleeding from radiation telangiectasia. Gastrointest Endosc 1997; 45:20-25.
6. Gupta N, Longo WE, Vernava AM III. Angiodysplasia of the lower gastrointestinal tract: an entity readily diagnosed by colonoscopy and primarily managed nonoperatively. Dis Colon Rectum 1995; 38:979-982.
7. Savides TJ, Jensen DM. Colonoscopic hemostasis for recurrent diverticular hemorrhage associated with a visible vessel: A report of three cases. Gastrointest Endosc 1994; 40:70-73.
8. Foutch PG, Zimmerman K. Diverticular bleeding and the pigment protuberance (sentinel clot): Clinical implications, histopathological correlation, and results of endoscopic intervention. Am J Gastroenterol 1996; 91:2589-2593.
9. Maunoury V, Brunetaud JM, Cortot A. Bipolar electrocoagulation treatment for hemorrhagic radiation injury of the lower digestive tract. Gastrointest Endosc 1991; 37:492-493.
10. Jensen DM, Jutabha R, Machicado GA et al. Prospective randomized comparative study of bipolar electrocoagulation versus heater probe for treatment of chronically bleeding internal hemorrhoids. Gastrointest Endosc 1997; 46:435-443.
11. Trowers EA, Ganga U, Rizk R et al. Endoscopic hemorrhoidal ligation: Preliminary clinical experience. Gastrointest Endosc 1998; 48:49-52.
12. Jensen DM, Machicado GA. Diagnosis and treatment of severe hematochezia: The role of urgent colonoscopy after purge. Gastroenterology 1988; 95:1569-1574.
13. Foutch PG. Angiodysplasia of the gastrointestinal tract. Am J Gastroenterol 1993; 88:807-818.

CHAPTER 12

Endoscopic Management of Variceal Bleeding

John S. Goff

Introduction

- Bleeding is one of the major complications of cirrhosis
  - Up to 40% of patients with variceal hemorrhage will expire within 6 weeks
  - 30-60% of the survivors will die within one year.
  - Bleeding is due to spontaneous rupture of an esophageal or gastric varix usually without a specific precipitating event.
- Control or prevention of bleeding could thus have a significant impact on survival and quality of life.
- The choices for treating acute variceal bleeding include:
  - Medications that lower portal pressure
  - Endoscopic sclerotherapy (EST)
  - Endoscopic variceal ligation (EVL)
  - Combination therapy
  - Endoscopic injection of glue (cyanoacrylate)
  - Sengstaken-Blakemore tube for tamponade of the bleeding site
  - Portal venous decompression with transjugular intrahepatic portal systemic shunt (TIPS) or surgical shunting
- Gastric varices pose a special problem for endoscopic treatment.
  - Treatment with EST, even when large volumes of sclerosant are used, has a very high rebleed rate.
  - Limited data suggests EVL combined with EST may be beneficial.
  - Glue injections have been reported to work well, but the agents are not available in the U.S.
- There are several choices for the prophylactic treatment of large varices.
  - EST – several studies have suggested worse outcome in the treated patients
  - EVL– no firm data yet, but extrapolating data from bleeding patients suggests this might work better than EST.
  - Medications have been shown to be successful in several studies.
- Beta-blockers (propranolol and others)
- Long acting nitrates (isosorbide)

Relevant Anatomy

- Normal anatomy
  - The superior mesenteric vein, inferior mesenteric vein and the splenic vein join to form the portal vein.
  - The left gastric vein (coronary vein) also flows into the portal vein.
  - About 1000 ml of blood flow through the portal vein each minute.

Gastrointestinal Endoscopy, edited by Jacques Van Dam and Richard C. K. Wong.

- Cirrhosis
  - Increased liver resistance leads to increased portal pressure.
  - Portal pressure greater than 12 mm Hg is associated with an increased potential for variceal bleeding.
  - Blood flow in the portal vein may actually flow away from the liver.
  - Splenic congestion causes enlargement and hypersplenism.
  - Esophageal and gastric varices develop because of increased pressure in the left gastric (coronary) vein, which flows from the esophagus and proximal stomach.

Indications for Endoscopic Therapy

- Visualization of active bleeding from a varix.
- Signs of recent bleeding from a varix – platelet plug.
- Large varices with no other explanation for a bleed.
- Prophylaxis in patients with large varices and no prior bleeding. Endoscopic therapy is controversial. Medical therapy is effective with beta-blockers or nitrates.

Contraindication to Endoscopic Therapy of Bleeding Varices

- Absolute contraindications
  - Unable to obtain informed consent.
  - Massive bleeding with no ability to visualize any landmarks in the proximal gastrointestinal tract.
  - Terminal patient.
- Relative contraindications. Encephalopathy or other reason for altered mental state
- Intubate patient to protect the airway.
- Consider paralyzing the patient. Very abnormal coagulation parameters—INR >2.5, platelets <20,000
- Transfuse with 3-6 units of fresh frozen plasma.
- Transfuse with 10 units of platelets.
- Consider DDAVP especially if uremic (0.2-0.4 mcg/kg IV)
  - Suspected or known poor long-term compliance
  - Gastric varices because of increased difficulty with control compared to esophageal varices.

Equipment

- Endoscopes
  - Start patient examination with large channel therapeutic endoscope when expecting blood in the stomach to allow for better aspiration of the gastric contents.
  - Use the standard endoscope for EVL (the device for this fits on the endoscope tip better).
- Devices
  - Needles for passing through the endoscope to do EST
  - Sclerosants for EST
- 5% ethanolamine oleate
- 1-3% sodium tetradecyl sulfate
- 5% sodium morrhuate
- Absolute ethanol

- Combinations. EVL devices.
- Single shot device
- Multiple ligator (4-10)
- Overtube needed for single shot; optional for multiple ligator.

Technique

- Medical therapy – preparing for endoscopy
- Gastric lavage – large bore orogastric tube (32 F or greater)
- Useful to assess amount of ongoing bleeding
- Clears stomach for better endoscopic viewing. Lower portal blood pressure.
- Intravenous vasopressin at 0.1-1 U/min (rarely use >0.6 U/min)
- Intravenous octreotide at 25-100 mcg/hr after a 50 mcg bolus
- Use nitroglycerin to offset side effects on blood pressure of the vasopressin (IV, PO or SL). Transfuse with red blood cells, platelets and fresh frozen plasma as needed.
- Endoscopic sclerotherapy (EST)
  - Freehand direct intravenous injection is the preferred technique.
  - Inject 0.5-2 ml per injection site, but may inject more than one site in a varix.
  - The injections need to be kept in the lower 5-10 cm of the esophagus.
  - The varices should have repeat injection in 7-10 days and then every 2 weeks until eradicated.
  - After eradication, endoscopy should be done at increasing intervals to monitor for return of varices.
- Endoscopic variceal ligation (EVL) (Fig. 12.1). Single shot versus multiple ligator.
- Overtube required for the single shot
- Multiple ligator results in a rather bulky endoscope tip
- Multiple ligator quicker . EVL can be done in any part of the esophagus.
- Usually treatment is in the lower 5-10 cm.
- One to 12 bands can be applied per session.
  - The timing for EVL sessions is the same as for EST.
  - Bands are applied by sucking the esophageal mucosa into the ligating device on the tip of the endoscope and then pulling a tripwire, which dispenses a band onto the varix. The band closes around the tissue with occlusion of the vein and subsequent clotting followed by tissue sloughing in a few days.
- Combination therapy (EVL and EST)
  - In the esophagus a single band is applied to each varix just above the gastroesophageal junction.
  - EST is then done above the ligation site with 1 ml of sclerosant per varix.
  - The other treatment parameters are the same as with EST or EVL alone.
  - In the stomach bands are applied to the varices just below the gastroesophageal junction usually with the endoscope in the retroflex view.
  - EST is then done more distally in the stomach with 2 ml per injection site.
- Endoscopic glue injection (cyanoacrylate)
  - There is limited data in the literature.
  - Gastric varices may be the best place for using this substance.
  - There is concern about systemic embolization of the glue.
  - Glue not available in the U.S. at this time.

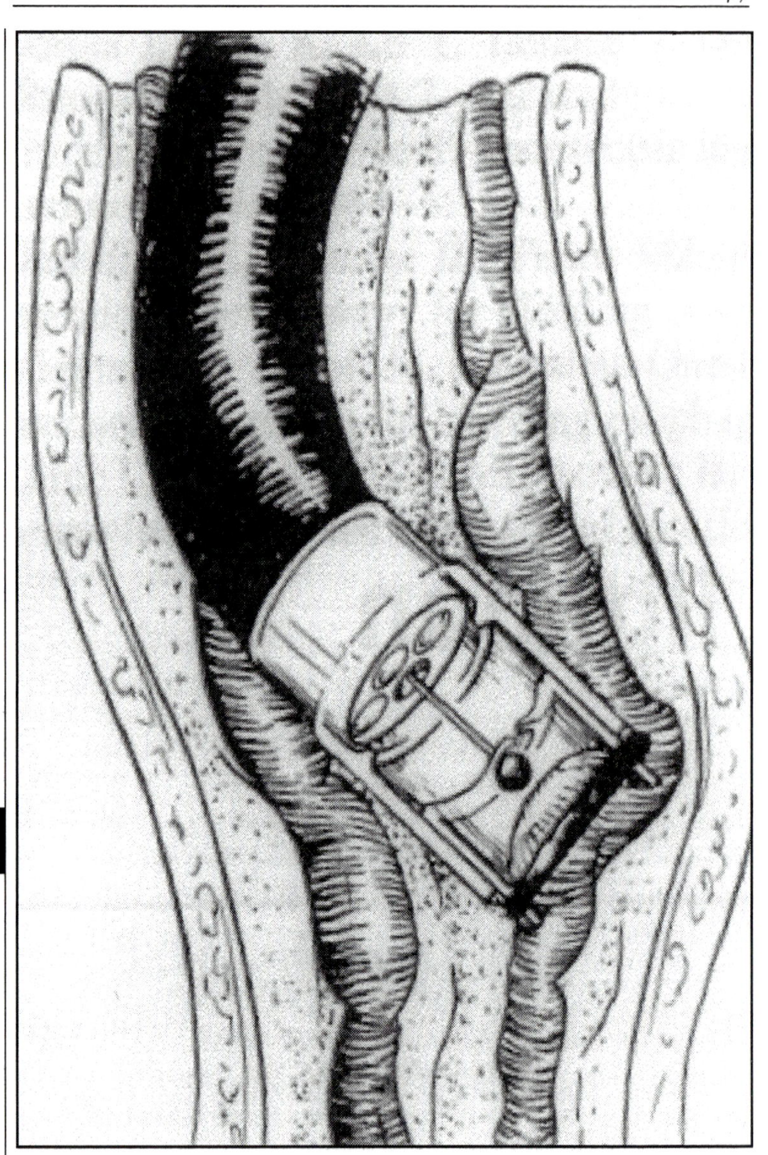

Figure 12.1. Endoscopic variceal ligation. The ligating device on the endoscope tip apposed to mucosa containing varix in the distal esophagus. Subsequently suction is applied and after aspirating the tissue into the device the band is deployed by pulling the trip wire.

Outcome of Endoscopic Therapy for Bleeding Varices

- EST. Interpretation of data confounded by different techniques/variables.
- Type and amount of sclerosant used
- Injection site – intravariceal or paravariceal
- Timing of retreatment
  - Active bleeding controlled in 80-90% of patients
  - Rebleeding rates significantly less than medically treated patients, especially if the varices are treated to obliteration. (average rebleed rate is 47%)
  - A survival advantage for those treated with EST compared to medical therapy has been hard to demonstrate in many trials. (25% fewer deaths based on meta-analysis)
  - Child's class C patients have the least benefit over the long-term when compared to A and B patients.
- EVL
  - Active bleeding controlled in about 90% of cases, but may require near blind placement of bands at the gastroesophageal junction initially if a lot of blood present. This is safer than blind injection therapy.
  - Rebleeding after successful EVL is similar to or less than with EST. (50% less rebleeding by meta-analysis)
  - A survival advantage compared to medical treatment and EST has been consistently demonstrated with EVL. (30% less deaths by meta-analysis)
  - There are many fewer complications with EVL.
  - Obliteration is usually achieved in one or more fewer sessions than with EST.
- Combination therapy
  - There is minimal advantage to using this method when treating esophageal varices.
  - This may be the endoscopic treatment of choice for gastric varices in the U.S., but the data is very limited currently.

Complication of Endoscopic Therapy for Bleeding Varices

- EST
- Serious complications
- Deep esophageal ulcers with perforation (1-6%)
- Pleural effusions
- Mediastinitis
- Strictures—most common complication (10-30%)
- Bacteremia (5-10%)
- Bleeding from esophageal ulcers. Minor complications.
- Chest pain (40%)
- Fever (40%)
- Dysphagia or odynophagia
- EVL. Serious complications.
- Proximal esophageal perforation from overtube use
- Esophageal strictures—very uncommon (<1%)
- Major bleeding—very uncommon (<1%). Minor complications.
- Dysphagia—transient
- Chest pain—uncommon

Selected References

1. Graham DY and Smith JL. The course of patients after variceal hemorrhage. Gastroenterology 1981; 80:800-805.
2. Lebrec D, De Fleury P, Rueff B et al. Portal hypertension, size of esophageal varices, and risk of gastrointestinal bleeding in alcoholic cirrhosis. Gastroenterology 1980; 79:1139-1146.
3. Chan LY and Sung JJY. Review article: The role of pharmaco-therapy for acute variceal haemorrhage in the era of endoscopic hemostasis. Aliment Pharmacol Ther 1997; 11:45-50.
4. Westaby D, MacDougall RD, Williams R. Improved survival following injection sclerotherapy for esophageal varices; Final analysis of a controlled trial. Hepatology 1985; 5:827-833.
5. Stiegmann GV, Goff JS, Michaletz-Onody PA et al. Endoscopic sclerotherapy as compared with endoscopic ligation for bleeding esophageal varices. NEJM 1992; 326:1527-1532.
6. Goff JS, Reveille RM, Stiegmann GV. Three years experience with endoscopic variceal ligation for treating bleeding varices. Endoscopy 1992; 24:401-404.
7. Ramond M-J, Valla D, Mosnier J-F et al. Successful endoscopic obturation of gastric varices with butyl cyanoacrylate. Hepatology 1989; 10:488-493.
8. Infante-Rivard C, Esnaola S, Villeneuve J-P. Role of endoscopic variceal sclerotherapy in long-term management of variceal bleeding: A meta-analysis. Gastroenterology 1989; 96:1087-1092.
9. Laine L, Cook D. Endoscopic ligation compared with sclerotherapy for treatment of esophageal variceal bleeding. Ann Intern Med 1995; 123:280-287.
10. Goff JS. Esophageal varices. Gastrointestinal Endoscopy Clinic of North America 1994; 4:747-771.

CHAPTER 13

Lasers in Endoscopy

Mark H. Mellow

Laser is an acronym for Light Amplification by Stimulated Emission of Radiation. Three properties combine to make a laser a powerful tool.

- Monochromicity—light in a laser beam is one the same wave length.
- Coherency—light waves in the beam are in phase with each other thus making amplitude additive.
- Highly directional—light travels in a narrow intense beam.

All laser effects are precise and tissue effect is predictable based on the physcial principle of that particular laser's effect on a target tissue. The reason different lasers have different biologic effects, irrespective of energy applied, is that they generate light of different wave length, which penetrates differently into biologic tissues. The Nd:YAG laser penetrates most deeply into tissue, approximately 4-5 mm; Argon laser and KTP lasers penetrate 1-2 mm; holmium YAG laser penetrates about 1 mm. These laser properties allow for selectivity in use. Destruction of neoplastic tissue is best accomplished with a deeply penetrating Nd:YAG; destruction of feeder vessels to large vascular malformations can be done with Nd:YAG also. In certain circumstances (thin-walled cecum, nerve-rich submucosa of the anal verge), KTP laser may be preferable.

Because of its versatility, the most common laser in endoscopic useis the Nd:YAG. A flexible laser wave guide is passed through the biopsy channel of an emdoscope and placed in close apposition to the target lesion (usually 0.5-1.0 cm). Energy is delivered via activation of a foot switch. Coaxial CO2 gas flows with the laser fiber, cools the tip and keeps the fiber free of debris. The YAG laser acts as intense heat source, coagulation occurs at a temperature of 60° C and vaporization at 100° C. The amount of energy delivered to a target is dependent upon the distance from fiber tip to target (inversely) and the power setting on the laser machine. For example, at a given distance, 40 watts employed for one second delivers less laser energy than 80 watts employed for one second. The YAG laser light is not visible to the human eye. Therefore a low power laser with a visible beam (helium neon) is coupled by mirrors to the therapeutic YAG laser beam. The YAG laser poses a definite hazard for eye damage. All personnel in the endoscopy suite *must* wear safety glasses during YAG laser use.

Photodynamic therapy (PDT) involves interaction of laser light with a sensitizing drug. The sensitizing drug must be present in the tumor or target tissue at a significantly higher concentration than elsewhere (selectivity). The drug should have fluorescent properties, should be cytotoxic after interacting with laser light and should be relatively nontoxic. Currently a hematoporphyrin derivative is used as the photosensitizing agent. In PDT the photosensitizing agent is give intravenously and is

Gastrointestinal Endoscopy, edited by Jacques Van Dam and Richard C. K. Wong.

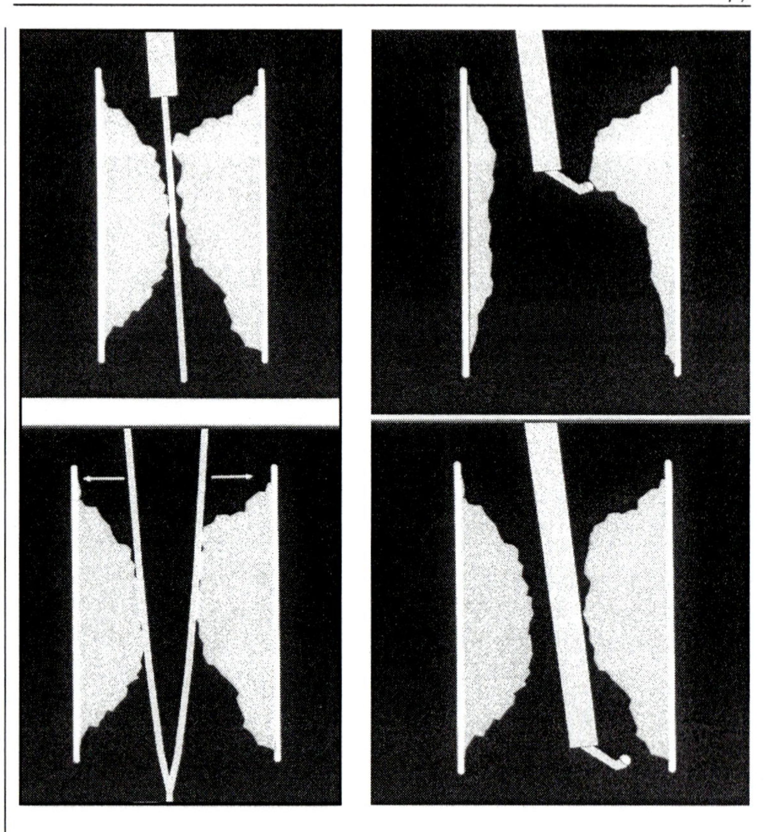

Figure 13.1. Treatment of esophageal CA. (top left) nearly occluding lesion; (bottom left) dilate so scope can be passed to distal tumor margin; (top right) treat margin first, work proximally; and (bottom right) treat proximal portion of tumor.

taken up by the target tissue relatively selectively. When activated by light of a suitable wave length (currently an argon dye laser), it produces singlet oxygen radicals in the cells which lead to cell destruction. A fiber emitting the required light is passed through the endoscope and placed in apposition to the tumor target. PDT is technically easier to deploy than Nd:YAG laser, which requires precise aiming, knowledge of depth of injury at a given power setting and target characteristics. Currently the drawback to PDT is significant dermatotoxicity. Hematoporphyrin derivative gets into the skin and produces burns when exposed to sunlight. As newer photosensitizing agents are developed, with higher specificity, less toxicity and can interact with light with great penetration potential, this modality will be of considerably more value. At present howeverm PDT has few clinical advantages over Nd:YAG laser.

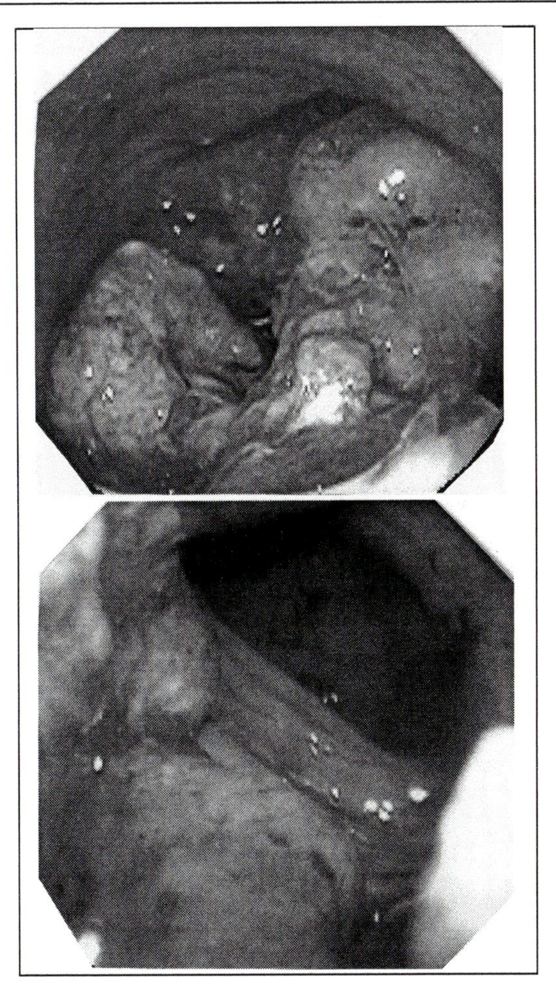

Figure 13.2. Rectal CA. (top) Before treatment and (bottom) after treatment.

Diseases Treated by Endoscopic Laser Therapy

Esophageal Carcinoma

First use of Nd:YAG laser in treatment of gastrointestinal neoplasms by Fleischer and Kessler, 1981.

- Indications. Primarily palliation of advanced disease or recurrent disease after initial chemotherapy and/or radiation therapy. In some circumstances, "up front" laser treatment followed by radiation therapy and chemotherapy (see below).

Table 13.1. Laser vs. stent for esophageal CA (covered metallic expandable is best stent)

	Laser	Stent
Repeated treatment required		X
Tumor overgrowth	X	
Stent migration retrieval	X	
Respiratory distress with esophageal stent (rare)	X	
TE fistula		X
Improvement in dysphagia grade		X

Table 13.2. Nd:YAG vs. PDT for esophageal CA

	Nd:YAG	PDT
Ease of treatment		X
Stricture formation	X	
Dermatotoxicity	X	
Difficulty with inflitrative or proximal lesions		X
Overall efficacy	X	X

Note: Neither laser nor stent is indicated in debilitated patients with anorexia and painful metastases. If any treatment considered, place PEG.

- Technique. Retrograde method. Dilate tumor first. Treat at distal tumor margein and move proximally. (Fig. 13.1) Coagulate and vaporize neoplasm so as to produce more patent lumen. Treat q2-3 days until patency achieved (average 2.5 sessions). Retreat prn dysphagia recurrence (mean approximately q10 weeks).
- Results. Luminal patency in 95% of patients; functional efficacy (ability to take necessary calories po to maintain nutritional status and leave hospital) in 70%. Proximal esophageal lesions and long lesions that traverse gastroesophageal junction are most difficult to treat. Polypoid lesions easier than infiltrative lesions. Perforation risk approximately 5%.
- Laser therapy as part of curative attempt. "Up front" laser therapy with radiation and chemotherapy to follow. Theory: debulk primary lesion, allow better luminal patency so that patients can ingest necessary calories better while undergoing subsequent treatments. 1-2 laser sessions just before start of radiation/chemotherapy.

Barrett's Esophagus

- Aim. Ablate entire area of Barrett's because leaving untreated tissue keeps patient at risk of carcinoma. Treat patients with dysplasia on biopsy. Use endoscopic therapy in conjunction with intensive antireflux therapy with hopes that normal squamous epithelium will regenerate over treated areas.

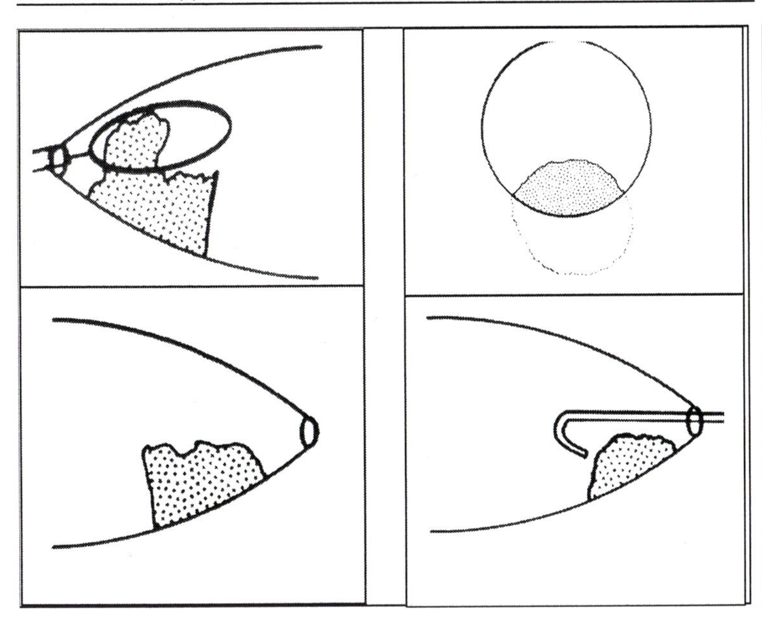

Figure 13.3. Treatment of rectal adenoma, polypoid and sessile components. (top left) remove polypoid portion via standard snare technique, specimen to Pathology; (bottom left) begin endoscopic laser treatment of sessile area; (top right) make sure area is examined in retroflexion because tumor often is not seen in end-on view; and (bottom left) treat, if necessary, in retroflexion and torque scope to get best angle; avoid scope shaft; may need to alter patient position.

- Photodynamic therapy. Largest clinical experience to date. Best technique for circumferential areas of Barrett's esophagus ≥ 3 cm in length. Small area of Barrett's, i.e., finger-like projections, < 3 cm circumferential involvement, can be treated with Nd:YAG laser.
- Risks. Perforation, stricture (common), Barrett's epithelium retained under new squamous mucosa.
- Bottom line. Still experimental. Patients with large areas of Barrett's with severe dysplasia who are acceptable risks should undergo surgery.

Early Gastric Carcinoma

Rare in the U.S. Small lesions, ≤ 3 cm. Japanese studies show good results with PDT and Nd:YAG used for superficial carcinoma (staged by endoscopic ultrasound) in poor operative risk patients. Raised lesions with discrete margins: Nd:YAG. Flat, ulcerated lesions with indiscrete margins: PDT. Complete remission at 2 years in 80%.

Advanced Gastric Carcinoma

YAG laser generally not very helpful. Minimal PDT experience; stents best for extensive antral lesions that cause partial gastric outlet obstruction.

Table 13.3. Complications of endoscopic laser therapy for colorectal cancers

	Number	%
Pain requiring narcotic analgesia	3	0.6
Bleeding requiring transfusion	10	2.0
≥ 10 ml	16	3.2
Perforation (local)	15	3.0
Perforation (free)	5	1.0
Anal stenosis	21	4.2
Fecal incontinence	5	1.0
Need for surgery	19	3.8
Death	6	1.2

*Survey by author involving 500 patients. (Contributors to the survey: Drs. Harvey Jacobs, Stephen Bown, Richard Dwyer, David Fleischer, Victor Grossier and Mark Mellow.)

Table 13.4. Endoscopic laser therapy for rectal and rectosigmoid villous adenomas and its relationship to the circumference of the tumor base

	Size of Tumor Base	
	Less than One-Third Circumference	More than Two-Thirds Circumference
Number of patients	114	39
Laser treatments per patient	3.2	13.5
Recurrence rate (%)	12	21
Stenosis needing dilation (%)	0	15
Subsequent carcinoma detection (%)	3	24

13

Table 13.5. Indications for endoscopic laser therapy for benign rectal and rectosigmoid adenomas in patients who are operative candidates and those who are not

Operative Candidate	Nonoperative Candidate
• Avoid interference with sphincteric function	• Control bleeding
• Recurrence following surgery	• Control of diarrhea
• "Certainty" of endoscopic cure	• Dehydration
• Relatively low cancer risk (size, appearance, sampling)	• Hypokalemia
• Patient refuses surgery	• Incontinence
• Alleviation of obstruction	

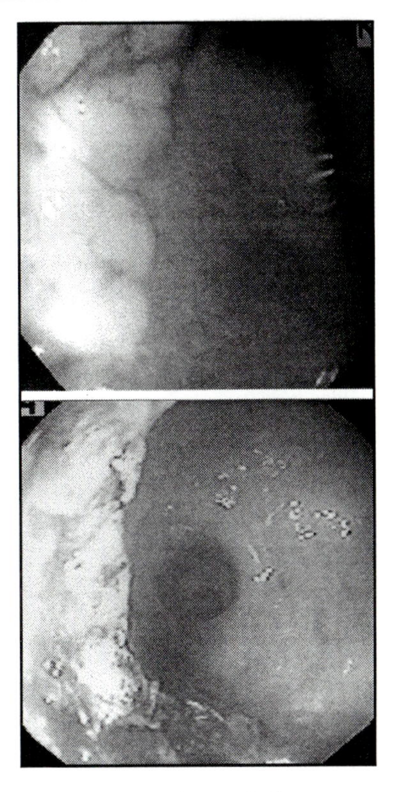

Figure 13.4a,b. Sessile rectal villous adenoma before treatment (top) and immediately after treatment (bottom).

Duodenal Malignancies

Laser occasionally of value. PDT better than Nd:YAG because of ease of treatment application. Palliation for bleeding and/or obstructive symptoms. Stents may be best for partially obstrcuting lesions. Difficult to treat this region with any modality.

Benign Duodenal Polyps

Isolated or with familial polyposis syndromes. Side viewing scope best. Lesions in ampulla common and, unfortunately, most difficult to ablate. Can invade or originate in distal CBD. May need Whipple operation.

Colorectal Cancer

- Background. Primary treatment of colorectal cancer is surgical, based on principles of relief and prevention of bleeding or obstruction and debulking primary tumor mass. However, approximately 10-15% of all patients with rectal cancer are better managed nonoperatively. This includes certain elderly patients, those with severe, associated medical conditions or with widespread

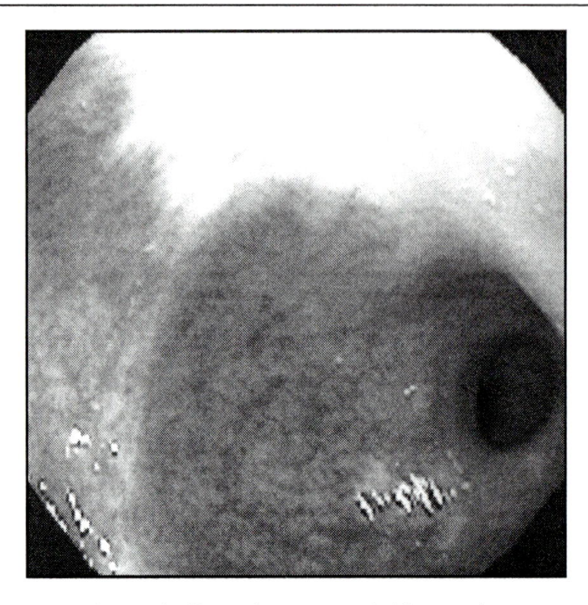

Figure 13.4c. Sessile rectal villous adenoma. Total ablation, after treatment.

metastases, and the occasional patient who refuses surgery. Rectal cancer is much more frequently treated with endoscopic laser therapy than more proximal colonic lesions because of: ease of access, need for more drastic surgery, less chance of severe complication (free perforation).

- Goal is usually palliative. Results: bleeidng controlled in 90%; obstruction managed in 75%.
- Treatment technique like that for esophageal cancer: IV sedation, coaxial CO2, Nd:YAG laser, 60-80 watts. Treat proximal tumor margin first (that portion furthest from anus) and work distally. Treat q2-4 days until lumen patency achieved or bleeding areas treated, usually accomplished in 2-3 sessions. (Fig. 13.2) Retreat approximately every 10 weeks. Most difficult to treat are circumferential lesions that traverse the rectosigmoid angle (stent preferable); lesions that extend to the anus, especially if circumferential; anastomotic recurrence since this is primarily extraluminal (stent preferable).
- The goal is cure in certain lesions, ie, ≤ 3 cm in length, ≤ one-thord of circumference, purely exophytic without ulceration. Brunetaud treated 19 such patients, 18 of whom had no local recurrence of clinically evident metastases at average follow-up of 37 months. However, since currently no fool-proof way to stage accurately to exclude nodal involvement (even with endoscopic ultrasound and/or radiolabeling with antitumor antibodies), patients still need at a relative surgicla contraindication to be considered for curative laser therapy forcolorectal cancer.
- Complications (Table 13.3)

Benign Colorectal Neoplasms

- Technique goal. Total ablation when possible, but ablation difficult with circumferential lesions and large circumferential lesions increase hidden carcinoma potential (Table 13.4). Therefore an otherwise healthy patioent with an extensive lesions should undergo surgery, even if it means AP resection with colostomy. The less extensive the lesion and the higher the surgical risk, the more appealing is laser treatment.
- Technique. (Fig. 13.3) Snare polypoid segments by standard snare polypectomy technique. Then treat remaining sessile tumor with laser. Purely sessile lesions are laser-treated (Fig. 13.4). In large lesions, obtain biopsy samples frequently to minimize the chance of missing carcinoma. For summary of indications for endoscopic laser therapy of benign rectosigmoid lesions see Table 13.5. With sessile villous adenomas, rationale for laser therapy versus surgery is similar, but lesions ≥ 4 cm (≥ one-half circumference) favor surgery.
- Alternative treatment. "Strip biopsy". Injection of normal saline into submucosa for "lif up" neoplasm, then polypectomy via standard snare technique. Good for smaller lesions. Very difficult for circumferential lesions and/or lesions near anal verge. Can inject saline into submucosa prior to laser treatment as well, which may decrease perforation risk.

Laser lithotripsy. Fragmentation of gallstones by laser. Larger common bile duct stones not amenable to removable via sphincterotomy or mechanical lithotripsy. Choledochoscope passed via biopsy channel of duodenoscope directly into common bile duct. Fiber placed in direct contact with stone. Not first-line therapy for choledocholithiasis.

Vascular Malformations of GI Tract

- Goal. Photocoagulation of mucosal vessels and, in some instances, submucosal feeder vessels.
- Indications for treatment. Severe iron deficiency anemia or acute bleeding. 50% present with occult bleeding; 50% with acute, sometimes recurrent, bleeding.
- Technique, Nd:YAG laser. 50-60 watts, 0.5-1,0 sec pulse duration. Aim to produce blanching of the surface yet coagulate down to submucosa. The mucosal lesion is often tip of the iceberg. Lesions that bleed spontaneously or ooze with initial treatment need the most aggressive therapy. Treat for at least 4 wk with acid suppressant therapy for iatrogenic, post-treatment ulcerations.
- Treatment pearls. Examine patient euvolemic. Hypovolemia cause blanching of vessels, hard to see. Water jet red spots, often AVMs, are mistaken for focal gastritis. If oozes blood with water jet, need to treat. Examine carefully on scope entry since scope trauma marks are easily confused with AVMs. Use glucagon or atropine to minimize gut motility on UGI exams. If barcotics are given for the procedure, reverse with Narcan; makes vessel easier to see. Check for coagulation defects, NSAID use, portal hypertension. These conditions *drastically* increase bleeding potential of AV malformations.
- Subgroups of AV malformations. Single, multiple, hereditary hemorrhagic telangiectasia; gastric antral vascular ectasia (GAVE, watermelon stomach).
- Results. Sustained reduction in transfusion requirements after laser treatment in approximately 75%. GAVE usually requires 2-3 treatments and sometimes repeat treatments at a later date. HHT patients most difficult to treat and may need adjunctive pharmacotherapy (estrogen-progesterone, danazol).

- Complications. Continued bleeding, not controlled (10%), perforation 2-3% in UGI tract and higher in right colon; antral narrowing, not clinically significant. Need for surgery, usually with GAVE, approximately 15%.

Radiation Proctitis

- Really a vasculopathy, after radiation to area of rectum (pelvic malignancies), resulting in chronic bleeding. Anemia is due to bleeding and radiation effect on pelvic bone marrow.
- Enema techniques rarely effective and can increase bleeding.
- Nd:YAG treatment
  - Treat as one treats AVMs. Coagulate surface vessel and submucosal feeder. Power 50-60 watts, 0.5-1.0 sec pulse duration at 1 cm distance to target. Anal verge most difficult.
  - Results. Decrease bleeding in many patients. However, sizeable minority of patients are not improved.
  - Complications. Perforation (3-5%); rectovaginal fistula; temporary increased bleeding from iatrogenic ulcer at treatment site.
- KTP laser. Similar goals and results to YAG. May be better at anal verge.

Selected References

1. Sargeant IR, Loizou LA, Rampton D et al. Laser ablation of upper gastrointestinal vascular ectasias: ;ong term results. Gut 1993; 34:470-475
2. Fleischer D, Sivak MV. Endoscopic Nd:YAG laser therapy as palliation for esophageal cancer—parameters affecting initial outcome. Gastroenterology 1985; 89:827-831
3. Sibille A, Descamps C, Jonard P et al. Endoscopic Nd:YAG treatment of superficial gastric carcinoma: Experience in 18 Western inoperable patients. Gastrointestinal Endosc 1995; 42:340-345
4. Brunetaud JM, Maunoury V, Cochelard D. Lasers in rectosigmoid tumors. Sem Surg Oncol 1995; 11:319-327
5. Laukka MA, Wand KK. Endoscopic Nd:YAG laser palliation of malignant duodenal tumors. Gastrointestinal Endosc 1995; 41:225-229
6. Overholt BF, Panjepour M. Photodynamic therapy in Barrett's esophagus: Reduction of specialized mucosa, ablation of dysplasia and treatment of superficial esophageal cancer. Sem Surg Oncol 1995; 11:372-376
7. Spinelli P, Mancini A, DalFante M. Endoscopic teatment of gastrointestinal tumors. Sem Surg Oncol 1995; 11:307-317
8. Mellow MH. Endoscopic laser therapy for colorectal neoplasms. Pract Gastroenterol 1997; 8:9-20
9. Van Cutsem E, Boonen A, Geboes K. Risk factors which determine the long term outcome of Nd:YAG laser palliation of colorectal cancer. Int J Colorect Dis 1989; 4:9-11
10. Mellow MH. Endoscopic laser therapy as an alternative to palliative surgery for adenocarcinoma of the rectun—Comparison of costs and complications. Gastrointestinal Endosc 1989; 35:283-287
11. Patrice T, Foultier MT, Yatayo S. Endoscopic photodynamic therapy with HPD for primary treatment of gastrointestinal neoplasms in inoperable patients. Dig Dis Sci 1990; 35:545-552
12. Brunetaud JM. Endoscopic laser treatment for a rectosigmoid villous adenoma: Factors affecting results. Gastroenterology 1989; 97:272-277
13. Barbatza C, Spencer GM, Thorpes M et al. Nd:YAG laser treatment for bleeidng from radiation proctitis. J Endosc 1996; 28:497-500

CHAPTER 14

Endoscopy of the Pregnant Patient

Laurence S. Bailen and Lori B. Olans

Introduction

Endoscopy is performed infrequently during pregnancy. Due to this limited number of procedures and the ethical considerations associated with clinical studies in gravid women, little published data are available on endoscopy in pregnancy. When considering upper or lower endoscopy in the gravid patient, special attention should be given to the diagnostic and therapeutic utility of the procedure and to the safety of the mother and fetus during the examination. Clearly, the threshold for performing endoscopy in the pregnant patient should be higher than in the non-pregnant patient given the potential risks. Nevertheless, in certain clinical situations endoscopy may be indicated to improve maternal and fetal well-being. This chapter will review the indications, techniques, findings, and outcomes of endoscopy performed during pregnancy.

Indications

- Endoscopy during pregnancy may play a role in making a definitive diagnosis in patients who are unresponsive to standard therapy or who have atypical symptoms. Endoscopy can also provide a safer alternative than more invasive approaches such as surgery.[1] Indications for esophagogastroduodenoscopy (EGD), flexible sigmoidoscopy, colonoscopy, and endoscopic retrograde cholangiopancreatography (ERCP) are considered below.

EGD[2]

- Upper gastrointestinal bleeding with hemodynamic compromise. Endoscopic therapeutic intervention may be possible.
- Symptoms of dysphagia, odynophagia, gastroesophageal reflux, nausea, vomiting, or abdominal pain which are severe, persistent, or refractory to empiric treatment.
- Suspected esophageal or gastric malignancy in which biopsy prior to postpartum period would influence management.

Flexible Sigmoidoscopy[2]

- Refractory distal colonic gastrointestinal bleeding (e.g., suspected colitis)
- Suspected rectal or sigmoid mass, stricture, or other obstructing lesion where biopsy prior to postpartum period would influence management.
- Severe refractory diarrhea of unclear etiology

Gastrointestinal Endoscopy, edited by Jacques Van Dam and Richard C. K. Wong.

Colonoscopy[2]

- Suspected proximal colonic malignancy or other mass of unclear etiology where biopsy prior to postpartum period would influence management.
- Severe, refractory bleeding due to proximal colonic source unreachable by flexible sigmoidoscope.

ERCP[2]

- Suspected refractory symptomatic choledocholithiasis, cholangitis, or gallstone pancreatitis.

Technique

- Special consideration should be given to medications used in preparation for (Table 14. 2) and during (Table 14.3) endoscopic examination of the pregnant patient. Noninvasive monitoring including blood pressure, pulse oximetry, and electrocardiography can aide in assessing the well-being of both mother and fetus.

Medication Safety

- The fetal safety of medications used during endoscopy is often determined by case reports in the medical literature and Food and Drug Administration (FDA) categorization.[1-4] Drugs are classified as category A, B, C, D, or X based on the level of risk to the fetus (Table 14.1).[4]

Preparation

- EGD and ERCP
  - Nothing to eat or drink for 8-12 hours prior to procedure.
  - Assure adequate hydration with intravenous fluids if necessary.
- Flexible sigmoidoscopy
  - Clear liquid diet day prior to examination.
  - Choices for distal bowel cleansing may include enemas, suppositories, and / or oral cathartics such as the following: Tap water or Fleet's enemas, dulcolax suppositories or tablets, and magnesium citrate oral solution. The FDA categorization for these medications regarding risk to the fetus is summarized in Table 14.2.
  - Safest options based on limited data: Gentle tap water enemas and/or dulcolax suppositories or tablets.[2]
- Colonoscopy (See Table 14.2)
  - Polyethylene glycol (PEG) solution (e.g., GoLytely, CoLyte, NuLytely). Patients must drink approximately 4 L of this isosmotic solution to achieve adequate bowel cleansing. No study on safety of PEG during pregnancy but limited data suggest safety when used in the puerperium.[5, 2]
  - Sodium phosphate solution (Fleet's PhosphoSoda). Patients often prefer this poorly absorbed salt solution which causes an osmotic diarrhea because the volume required for bowel cleansing is less than the volume of PEG solution needed. Little data on safety available. One case report of bone growth failure in an infant born to an anorexic mother with maternal phosphate overload due to excessive phosphate enema use during pregnancy.[6, 2]

Table 14.1. FDA categorization based on fetal effects

Category A	Controlled studies in women fail to demonstrate a risk to the fetus in the first trimester (and there is no evidence of a risk in later trimesters), and the possibility of fetal harm appears remote.
Category B	Either animal-reproduction studies have not demonstrated a fetal risk, but there are no controlled studies in pregnant women or animal-reproduction studies have shown an adverse effect (other than a decrease in fertility) that was not confirmed in controlled studies in women in the first trimester (and there is no evidence of a risk in later trimesters).
Category C	Either studies in animals revealed adverse effects on the fetus (teratogenic or embryocidal, or other) and there are no controlled studies in women or studies in women and animals are not available. Drugs should be given only if the potential benefit justifies the potential risk to the fetus.
Category D	There is positive evidence of human fetal risk, but the benefits from use in pregnant women may be acceptable despite the risk (e.g., if the drug is needed in a life-threatening situation or for a serious disease for which safer drugs cannot be used or are ineffective).
Category X	Studies in animals or human beings have demonstrated fetal abnormalities, or there is evidence of fetal risk based on human experience, or both, and the risk of the use of the drug in pregnant women clearly outweighs any possible benefit. The drug is contraindicated in women who are or may become pregnant.

Drugs Used as Premedications and During Endoscopy

- Flexible sigmoidoscopy is routinely performed without premedication. However, given the discomfort associated with upper endoscopic and colonoscopic procedures, only rare patients are able to complete these procedures without medications.
- Table 14.3 outlines common medications used during endoscopy to enhance patient comfort along with the associated FDA categorization.
- Diazepam is a benzodiazepine which may cause neonatal floppy infant syndrome (hypotonia, lethargy, irritability) if given to mothers during labor.[4, 2] There is a suggested but not proved increased risk of congenital malformations and central nervous system problems when given to pregnant women.[2, 4]
- Midazolam is a newer benzodiazepine compared to diazepam. Less data on its use in pregnancy are available. Midazolam prior to caesarean section may have a depressant effect on newborns.[4 2] In many reports of use during endoscopy in pregnancy, midazolam caused no obvious illeffects.[79]
- Meperidine is commonly used for endoscopic premedication during pregnancy. No known fetal problems during pregnancy except when given during labor when it may cause transient respiratory depression and impaired alertness.[4, 2]
- Fentanyl is a Category B drug with no known associated congenital defects. One case of respiratory depression in an infant born to a mother who received epidural fentanyl during labor.[4, 2]

Table 14.2 Bowel preparation prior to colonoscopy in the pregnant patient

Medication	Category
Polyethylene glycol (GoLytely, CoLyte, NuLytely)	Category C
Magnesium citrate	Unlabeled
Sodium phosphate solution (Fleets Phospho-Soda, Fleet's enema)	Unlabeled
Dulcolax suppositories/tablets	Category B

- Droperidol is a butyrophenone derivative with sedative and antiemetic effects. It has been used as an adjunct to sedation in caesarean sections and in the management of hyperemesis gravidarum without documented fetal harm.[4]
- Simethicone is a silicon product which eliminates gas bubbles that may impair endoscopic visualization. Use of this Category C drug during pregnancy is usually avoided due to limited data reporting a possible increase in birth defects.[4, 2]
- When necessary, medications administered during pregnancy should be given judiciously due to the lack of definitive studies regarding fetal outcome. Based on the available data, meperidine and fentanyl are likely safer medications when compared with diazepam and midazolam.[1]

Monitoring

- Noninvasive monitoring provides valuable information prior to and during endoscopic procedures to assist in maximizing maternal and fetal well-being.
- Recommended monitoring includes the following:[2]
- Supplemental oxygen with continuous pulse oximetry
- Blood pressure and telemetry monitoring.
- Continuous fetal heart monitoring or, when not technically possible, intermittent fetal monitoring.
- Anesthesia support may be helpful for long or complicated procedures to assist with medication administration and monitoring.
- Abdominal pelvic shielding with lead should be used when fluoroscopy is needed during ERCP. Procedures should be performed by experienced endoscopists in order to minimize fluoroscopy time.

Results and Outcomes of Endoscopy During Pregnancy

- Esophagogastroduodenoscopy
  - The largest series of EGD during pregnancy, a case control study, included 83 procedures.[8]
  - Gastrointestinal bleeding was the most common indication.
  - The most frequent finding for this indication was reflux esophagitis. Other findings included Mallory-Weiss tear, gastritis, and duodenal ulcer.
  - Common findings for symptoms of nausea, vomiting, and abdominal pain were esophagitis and gastritis.
  - Meperidine, midazolam, diazepam, and naloxone were all used without incident.

Table 14.3 Medication for GI endoscopy in the pregnant patient

Medication	Category
Diazepam	D
Midazolam	D (categorized by Briggs [1])
Meperidine	B
Fentanyl	B
Droperidol	C (categorized by Briggs)
Simethicone	C (categorized by Briggs)
Glucagon	B
Lidocaine 10% oral spray	B
Flumazenil	C
Naloxone	B
Atropine	C

[1] Briggs GG, Freeman RK, Yaffe SJ. Drugs in pregnancy and lactation : A reference guide to fetal and neonatal risk. 4th ed. Baltimore: Williams & Wilkins, 1994.

  - No maternal complications, no induction of labor, no congenital malformations, and no poor fetal outcomes were reported directly related to endoscopy.
  - There are scant data regarding endoscopic hemostasis in pregnancy.[2] The benefits and potential risks of the procedure must be compared with the benefits and risks of other therapeutic modalities such as surgery.
- Flexible sigmoidoscopy
  - The largest study of pregnant women undergoing flexible sigmoidoscopy included 46 patients. [7] Procedures were performed throughout all trimesters of gestation.
  - The most common indication was hematochezia. This indication also had the highest diagnostic yield.
  - Inflammatory bowel disease and internal hemorrhoids were the most common findings. Other findings included infectious colitis, polyps, anastomotic ulceration, colon cancer, and normal examination.
  - No maternal complications were noted. No significant differences in fetal outcome (including congenital defects, premature delivery, and Apgar scores) compared to control group and to national averages. No poor fetal outcomes (including stillbirth and involuntary abortion) could be ascribed to the procedure.
- Colonoscopy
  - Limited data on colonoscopy during pregnancy.
  - Largest published series of colonoscopy during pregnancy included eight patients. Four procedures were performed in first trimester and four were performed in the second trimester.[7]
  - Indications for colonoscopy included persistent bloody diarrhea and abdominal pain.
  - Most common finding was colitis. Other findings included hemorrhoids, anastomotic ulceration, and normal examination.
  - No reported fetal or maternal complications related to the procedure.

- Endoscopic retrograde cholangiopancreatography
  - The largest published series of ERCP during pregnancy included 29 procedures performed throughout gestation.[9]
  - Indications for ERCP included abdominal pain with abnormal liver function tests with or without abnormal abdominal ultrasound.
  - The most common finding was choledocholithiasis. Other findings included primary sclerosing cholangitis (PSC), biliary leak, pancreas divisum, pancreatic duct stricture, and normal examination.
  - Therapeutic procedures included endoscopic sphincterotomy, biliary stent placement, stone extraction, and stricture dilatation.
  - Meperidine, diazepam, midazolam, glucagon, and atropine were used without incident.
  - ERCP was performed in the prone position or in the left lateral decubitus position in later stages of pregnancy. Fluoroscopy time was minimized by directly cannulating with a sphincterotome and aspirating bile to confirm location.
  - Pancreatitis occurred following each of three ERCPs in one pregnant patient. One spontaneous abortion and one neonatal death were reported without apparent causal relationship to the procedure.
  - The present role of endoscopic ultrasound and magnetic resonance cholangiopancreatography (MRCP) in pregnant patients with pancreaticobiliary symptoms and signs is unknown. Future studies may demonstrate less maternal and fetal risks with these procedures.[2]

Conclusion

- Endoscopic procedures during gestation can provide valuable diagnostic information, offer definitive therapy, and avoid more invasive approaches such as surgery. The little data available suggest that endoscopy can be performed safely in the pregnant patient. However, given the potential risks endoscopy should only be performed when clearly indicated. Careful attention should be given to the safety of medications used and to noninvasive monitoring in order to maximize maternal and fetal well-being. Close collaboration between gastroenterologist and obstetrician is essential.

Selected References

1. Olans LB, Wolf JL. Gastroesophageal reflux in pregnancy. Gastrointest Endosc Clin N Am 1994; 4:699-712.
2. Cappell MS. The safety and efficacy of gastrointestinal endoscopy during pregnancy. Gastroenterol Clin North Am 1998; 27:37-71.
3. Koren G, Pastuszak A, Ito S. Drugs in pregnancy. N Engl J Med 1998; 338:1128-1137.
4. Briggs GG, Freeman RK, Yaffe SJ. Drugs in pregnancy and lactation : A reference guide to fetal and neonatal risk. 4th ed. Baltimore: Williams & Wilkins, 1994.
5. Nardulli G, Limongi F, Sue G et al. Use of polyethylene glycol in the treatment of puerperal constipation. G E N 1995; 49:224-226.
6. Rimensberger P, Schubiger G, Willi U. Connatal rickets following repeated administration of phosphate enemas in pregnancy: A case report. Eur J Pediatr 1992; 151:54-56.

7. Cappell MS, Colon VJ, Sidhom OA. A study at 10 medical centers of the safety and efficacy of 48 flexible sigmoidoscopies and 8 colonoscopies during pregnancy with follow-up of fetal outcome and with comparison to control groups. Dig Dis Sci 1996; 41:235-361.
8. Cappell MS, Colon VJ, Sidhom OA. A study of eight medical centers of the safety and clinical efficacy of esophagogastroduodenoscopy in 83 pregnant females with followup of fetal outcome with comparison control groups. Am J Gastroenterol 1996; 91:348-354.
9. Jamidar PA, Beck GJ, Hoffman BJ et al. Endoscopic retrograde cholangiopancreatography in pregnancy. Am J Gastroenterol 1995; 90:126-137.

CHAPTER 15

Percutaneous Endoscopic Gastrostomy and Percutaneous Endoscopic Jejunostomy

Richard C. K. Wong and Jeffrey L. Ponsky

Introduction

It has been two decades since the original description by Gauderer, Ponsky and Izant of the endoscopically guided technique for creating a tube gastrostomy. Since then, the placement of a percutaneous endoscopic gastrostomy (PEG) tube has been successfully used in children and infants, adults and in the elderly. In particular, it has become a common long-term route of providing enteral nutrition to patients who are unable to maintain sufficient oral intake.

Indications

- To provide nutrition to those patients who have a functional gastrointestinal tract but yet are unable or unwilling to sustain sufficient oral intake to keep up with their caloric needs.
- PEG tubes should only be used in patients who are thought to require long-term enteral nutrition (i.e., for > 30 days). In the short-term (i.e., < 30 days), small bore nasogastric or nasoenteric tubes are strongly recommended.
- Patients requiring PEG placement commonly have severe neurological impairment to swallowing function. Alternatively, there may be mechanical obstruction of the upper aerodigestive tract, severe facial trauma, following maxillofacial tumor surgery or in patients with developmental disorders affecting the mechanism of swallowing.
- Other indications are outlined in Table 15.1.

Contraindications

- Patients who are unlikely to benefit from it or in patients who are not expected to live for any significant length of time.
- Patients in whom the anterior abdominal wall cannot be brought into close proximity with the anterior gastric wall (e.g., significant ascites, morbid obesity, prominent enlargement of left lobe of liver or splenomegaly).
- Nonfunctional gastrointestinal tract or uncorrectable coagulation abnormalities and the presence of gastric varices.
- Relative contraindications include previous gastric or major abdominal surgery, the presence of a ventriculoperitoneal shunt and recent myocardial infarction (Table 15.2).

Technique

- Patient preparation

Gastrointestinal Endoscopy, edited by Jacques Van Dam and Richard C. K. Wong.

Table 15.1. Indications for PEG placement

Longterm enteral nutrition (> 30 days)
Neurologic or developmental impairment
Mechanical obstruction of the upper aerodigestive tract
Severe facial trauma
Following maxillofacial surgery
Gastric decompression
Gastric atony
Carcinomatosis
Recalcitrant intestinal obstruction
Other:
Chronic administration of unpalatable medications or diets
Conduit for bile replacement in patients with external biliary fistula
Gasbloat syndrome
Fixation of recurrent gastric volvulus

- Informed consent is obtained from the patient or from the appropriate surrogate if the patient is mentally incompetent. This should follow a detailed discussion with the patient and their family of the indications, risks/complications, benefits and alternative methods of feeding or decompression.
- Feedings are withheld from the patient for eight hours prior to the procedure. A single prophylactic dose of an antibiotic, usually a cephalosporin, is administered intravenously just prior to the procedure.

The "Pull" Method

- Standard esophagogastroduodenoscopy is performed.
- The stomach is fully inflated with air to displace adjacent organs and to approximate the anterior walls of both the stomach and the abdomen.
- The endoscopist must be able to clearly see the transilluminated light from the endoscope through the anterior abdominal wall and localize the site of indentation (by external finger pressure) on the anterior gastric wall.
- The selected site on the anterior abdominal wall is then prepared and draped in a sterile fashion. Local anesthetic is applied and a transverse skin incision of approximately 1 cm is made.
- The stomach is again fully inflated with air and a puncturing cannula is thrust through the incision into the gastric lumen. An endoscopic snare is tightened around the cannula and the inner stylet removed. A loop of suture or wire is passed through the cannula into the stomach and the snare readjusted and firmly tightened around this material (Fig. 15.1). The suture or wire loop is pulled, with the endoscope, from the patient's mouth.
- The suture or wire loop is then firmly affixed to the dilating end of a well-lubricated PEG tube and the entire complex is pulled by external traction (hence, "Pull" technique) from the abdominal end of the suture, down through the esophagus and into the stomach (Fig. 15.2).
- The anchoring outer crossbar (otherwise known as an external bumper) is then applied and the distance (in centimeters) between the crossbar and the skin surface is documented in the procedure report (Fig. 15.3). It is important to insure that the crossbar is not positioned too tightly as this can lead to local

Table 15.2. Contraindications to PEG placement for purposes of enteral feeding

Absolute:
- -patients not expected to live for any significant length of time
- -patients who are unlikely to benefit from enteral feedings
- -inability to obtain informed consent
- -inability to approximate anterior abdominal and gastric walls
- -nonfunctioning or obstructed gastrointestinal tract
- -uncorrectable coagulopathy
- -gastric varices
- -chronic peritoneal dialysis
- -absence of stomach

Relative:
- -previous gastric or major abdominal surgery
- -ventriculoperitoneal shunt
- -recent myocardial infarction

tissue ischemia with subsequent ulceration/pain, infection, abscess formation and the "buried bumper syndrome". A distance of several millimeters is essential between the skin and the outer crossbar.

- Topical antibiotic ointment is applied to the PEG site and covered by a light (non-occlusive) sterile dressing. The patient (or their caregiver) is instructed not to use the PEG tube for approximately 24 hours.

The "Push" Method

- This technical modification affords results which are essentially similar to those of the "pull" method. Instead of a flexible suture or wire loop, a much stiffer wire is used to act as a guide upon which the rigid dilator end of the PEG tube is threaded. Both ends of the guide-wire are held taut as the PEG tube is pushed (hence, "Push" technique) over the wire, into the stomach and then out of the anterior abdominal wall. The dilator end of the PEG tube is then grasped and pulled to its final position.

The "Single-Step Button" Method

- The gastrostomy button is a skin-level, continent gastrostomy device of variable shaft length. Thus far, its primarily use has been as a PEG tube replacement device after a mature gastrocutaneous tract has formed (hence, a two-stage procedure). Since then, it has been further developed and marketed for use as a single or one-step procedure.

Feeding and Local Care

- Examine the patient prior to tube feeding.
- Examine the abdomen and the gastrostomy tube insertion site closely for signs of infection. The outer crossbar should be loosened if it is too tight: this is particularly important as excessive traction increases the risk of local complications. An occlusive dressing is not recommended but a simple dressing can be applied to the tube site daily for the first few days. Thereafter, the tube site can be left uncovered and diluted hydrogen peroxide (on a cotton swab) can be used to periodically remove the buildup of debris around the insertion site.

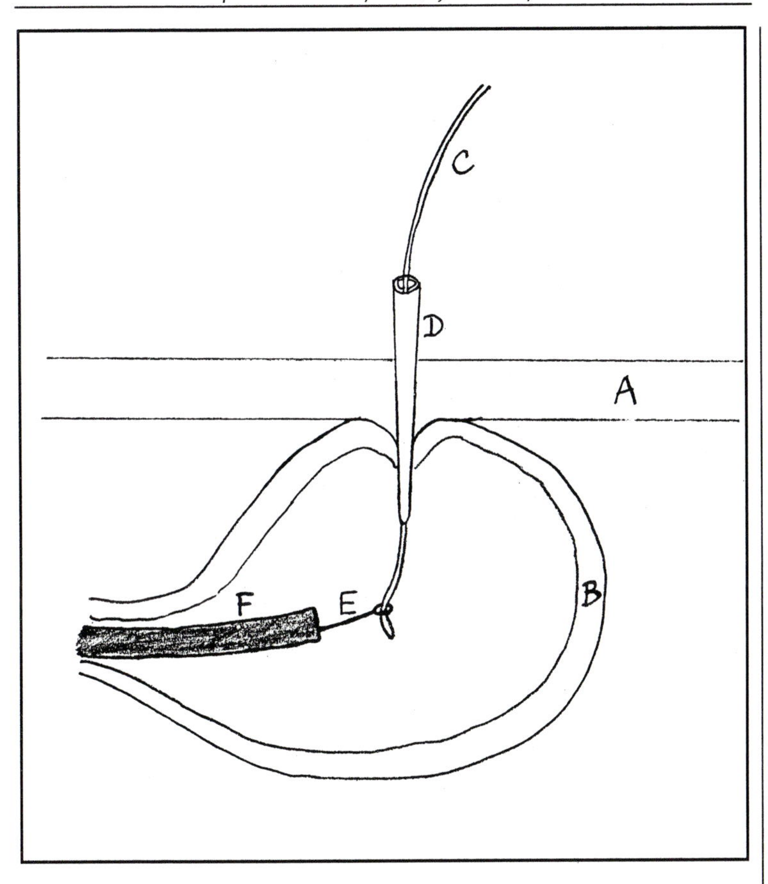

Figure 15.1. An endoscopic snare is used to firmly grasp the wire loop (or suture) which has been inserted through the puncturing cannula into the gastric lumen: (A) abdominal wall, (B) gastric wall, (C) wire loop (or suture), (D) cannula, (E) snare and (F) endoscope.

Percutaneous Endoscopic Gastrostomy Tube Replacement

- Once the gastrostomy tube has been removed (intentionally or inadvertently) the gastrocutaneous tract closes quickly (usually well within 24 h). A mature gastrocutaneous tract forms within a few weeks following PEG placement.
- With a mature tract, no special management is needed apart from a simple sterile dressing and topical antibiotic application for a few days. However, if the tube is inadvertently removed and replacement is desired (in a mature tract), a Foley catheter should be immediately inserted in order to maintain tract patency. A more permanent gastrostomy tube can be inserted nonendoscopically on an elective basis. In an immature tract the patient should be carefully monitored for signs of acute peritonitis. If the patient is clinically stable, the gastrostomy tube should be replaced endoscopically.

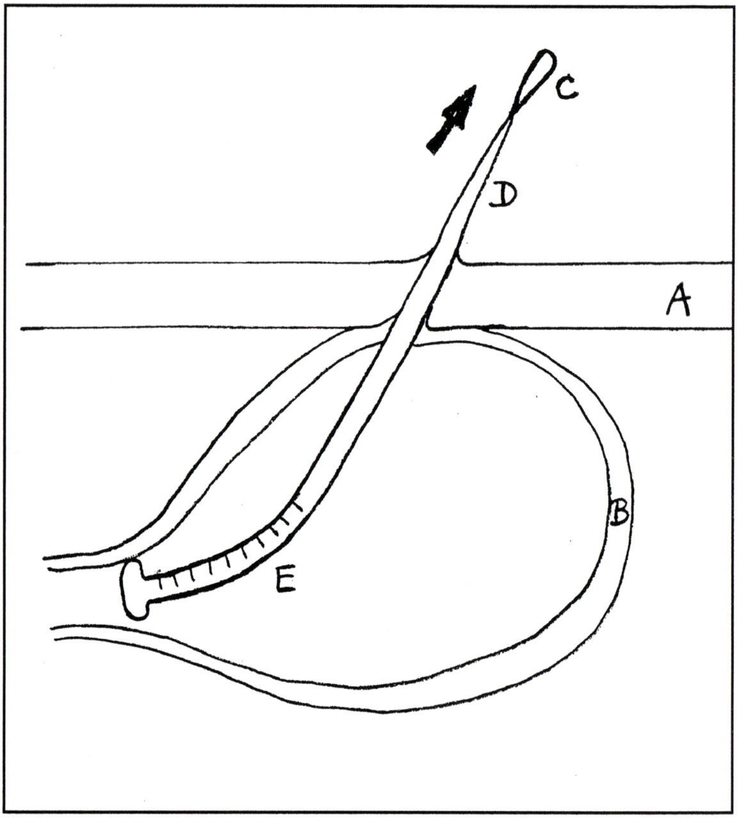

Figure 15.2. The "pull" method: after fixation of the wire loop (or suture) to the dilating end of the gastrostomy tube, the entire complex is pulled by external traction down through the esophagus and into the stomach: (A) abdominal wall, (B) gastric wall, (C) wire loop at..., (D) dilating end of gastrostomy tube, (E) intragastric bumper end of gastrostomy tube.

Complications

- Careful adherence to sterile technique and the proper administration of prophylactic antibiotics are essential to decrease infectious complications.
- *Local cellulitis/abscess*—may present as erythema, pain with tenderness and purulent discharge around the PEG tube insertion site. Local fluctuance may indicate the presence of a peritubal abscess. Appropriate antibiotics should immediately be instituted and local drainage of any subcutaneous collection of pus performed. Occasionally, local fungal infection may occur which will require the application of topical antifungal preparations.
- *Necrotizing fasciitis*—This rare, but potentially deadly spreading infection of the abdominal wall requires immediate intravenous antibiotics and surgical consultation, as radical surgical debridement may be necessary.

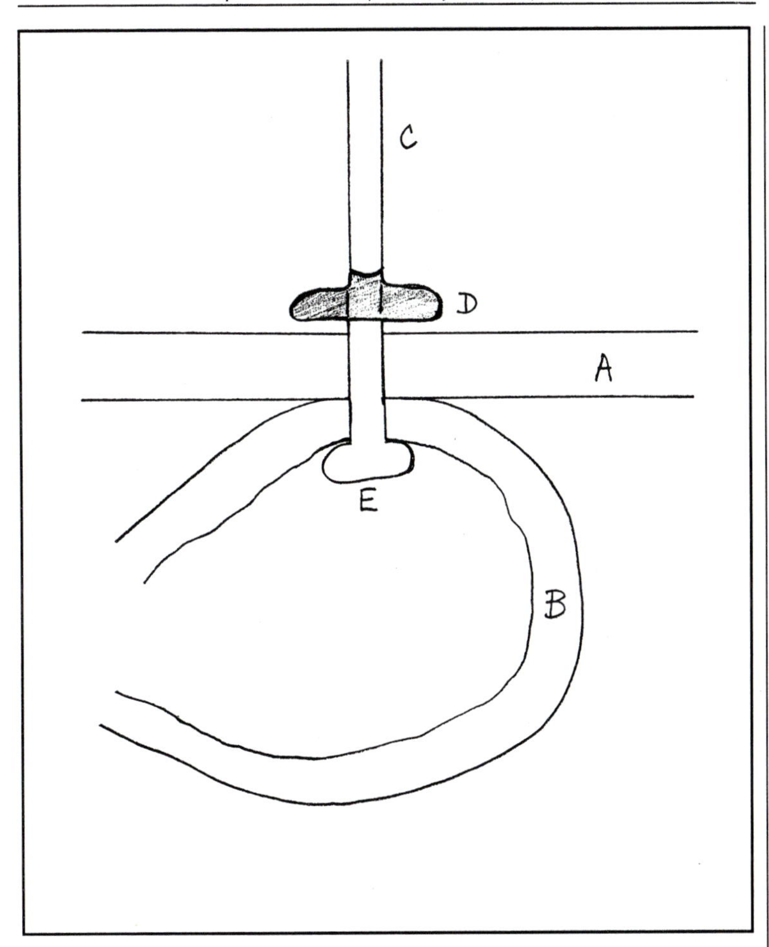

Figure 15.3. Final position of the gastrostomy tube: (A) abdominal wall, (B) gastric wall, (C) gastrostomy tube, (D) outer crossbar (external bumper), (E) intragastric bumper.

Tube Migration

- *Proximal*—Tube migration externally into the abdominal wall is termed the, "buried bumper syndrome." This usually occurs because the outer crossbar has been positioned too tightly causing excessive tension to the bumper. Presenting features may include increased resistance to flow of liquid through the tube, erythema, pain and tenderness around the tube insertion site. When this complication occurs, the intragastric bumper will need to be loosened or repositioned entirely depending on the degree of external migration. This can be accomplished by both surgical and nonsurgical techniques.

Table 15.3. Complications of PEG placement

- Infectious (local cellulitis, abscess, necrotizing fasciitis) - Tube migration (proximal, distal) - Separation of stomach from abdominal wall - Fistulae (gastrocolic, colocutaneous) - Enlargement of gastrostomy site - Implantation metastases - Others: small bowel fistula, intestinal volvulus, colonic and duodenal obstruction, pyoderma gangrenosum, acute gastric dilatation.

- *Distal*—Unintentional loosening of the outer crossbar can occur with repeated manipulation (e.g., cleaning) of the insertion site. This can lead to the distal migration of the intragastric bumper (or balloon) and result in gastric outlet or duodenal obstruction. It is important to keep the outer crossbar at the same distance on the gastrostomy tube after each manipulation.

Fistulae

- *Gastrocolic fistula*—This complication may present late and is due to the penetration of the colon by the gastrostomy tube at time of placement.
- *Colocutaneous fistula*—This complication is rare and occurs following penetration of the colon and the external migration of the entire intragastric bumper from the stomach to the colon.

Enlargement of the Gastrostomy Insertion Site

- This results from either excessive tension or pivoting of the tube as it exits the abdominal wall. It can lead to leakage of gastric contents onto the skin with all its attendant problems. The solution is to either replace the tube with a skin level device (such as a button) or replace the gastrostomy tube at a new site.

Implantation Metastases

- This is a rare occurrence and has been reported in patients with head and neck carcinoma. The exact pathologic mechanism is unknown but postulated theories have included the direct transfer of tumor cells from the obstructing neoplasm to the tube insertion site, and tumor seeding via the hematogenous route.

Radiologic Pneumoperitoneum

- This is not an actual complication. In fact, it has been reported in approximately 40% of cases following PEG placement. In the vast majority of instances, the pneumoperitoneum is clinically benign and is without clinical consequence. Therefore, routine ordering of plain abdominal radiographs following uneventful PEG placement cannot be justified.

Comparing Surgical, Endoscopic and Radiologic Gastrostomy

- Endoscopic placement is more cost-effective and less invasive than standard surgical gastrostomy. It is also preferable to radiologic gastrostomy as it permits a thorough examination of the upper digestive tract which may detect occult (but significant) pathology. In general, endoscopic placement should be attempted first unless there are specific contraindications to esophagogastroduodenoscopy.

Specific indications for a surgical gastrostomy include instances where endoscopic placement is not possible (e.g., previous major abdominal surgery).

Percutaneous Endoscopic Jejunostomy

- The major indication for the placement of a jejunostomy feeding tube is in patients with documented aspiration of tube feeds unresponsive to standard medical measures (e.g., elevation of head of bed, lower rate of tube feeding, prokinetic medications).
- Surgically placed jejunostomy feeding tubes have performed well in this regard. However, endoscopic placement has been problematic and less successful. This has been primarily due to the fact that endoscopically placed tubes often migrate proximally back into the duodenum or even into the stomach. Future technical developments (e.g., *direct* percutaneous endoscopic jejunostomy placement using push enteroscopy technique) may permit the endoscopic placement of jejunostomy tubes which remain in the jejunum.

Ethical Considerations

- Placement of a PEG tube for feeding has become commonplace. It is technically straightforward and can be accomplished in most patients. The critical issue is the selection of patients who will obtain significant benefit from gastrostomy tube placement. PEG placement may not be appropriate in some patients in whom improvement in functional or nutritional status, or overall life expectancy is not anticipated. In such patients, nasoenteric tube feedings should be used, if appropriate. Thoughtful consultation with the patient and family is essential and should precede the placement of a PEG tube in all cases.

Selected References

1. Gauderer MW, Ponsky JL, Izant RJ Jr. Gastrostomy without laparotomy: A percutaneous endoscopic technique. J Pediatr Surg 1980; 15:872-875.
   Original description of the PEG technique.
2. Wong RCK, Ponsky JL. Percutaneous endoscopic gastrostomy. In: Sivak MV Jr, ed. Gastroenterologic Endoscopy, 2nd ed. Philadelphia: WB Saunders Co. (in press).
   Comprehensive discussion and review of PEG/PEJ.
3. Ponsky JL. Percutaneous endoscopic gastrostomy: techniques of removal and replacement. Gastrointest Endosc Clin N Am 1992; 2:215-221.
   This article focuses on the indications for and methods of replacing gastrostomy devices.
4. Foutch PG, Talbert GA, Waring JP et al. Percutaneous endoscopic gastrostomy in patients with prior abdominal surgery: Virtues of the safe tract. Am J Gastroenterol 1988; 83:147-150.
   Describes the use of the "Safe Tract" technique.
5. Shike M, Latkany L, Gerdes H et al. Direct percutaneous endoscopic jejunostomies for enteral feeding. Gastrointest Endosc 1996; 44:536-540.
   This article describes the technique of single-step, direct PEJ placement.
6. American Gastroenterological Association Medical Position Statement: Guidelines for use of enteral nutrition. Gastroenterology 1995; 108:1280-1281.
   Position Statement by the AGA on use of enteral nutrition.
7. American Gastroenterological Association technical review on tube feeding for enteral nutrition. Gastroenterology 1995; 108:1282-1301.
   Detailed review on the technical aspects of enteral tube feeding.

CHAPTER 16

Small Bowel Endoscopy

Jeffery S. Cooley and David R. Cave

Introduction

Standard upper and lower gastrointestinal endoscopy effectively evaluates a variety of pathologic conditions affecting the esophagus, stomach, proximal duodenum and colon respectively. However, the distal duodenum, jejunum and ileum are left unexamined by these procedures. Conditions such as chronic gastrointestinal bleeding, malabsorption or abnormalities found by the use of contrast studies may require the evaluation of the small intestinal mucosa. Many approaches have been used to investigate the small bowel including standard barium studies, enteroclysis, angiography, bleeding scans and surgery. The length of the small intestine, it's flexibility, small lumen and coiled nature have made visualization of the small intestinal mucosa by endoscopic means a challenging and often unsatisfactory undertaking. Enteroscopy refers to specific endoscopic techniques with specialized equipment that permits more thorough small bowel evaluation than can be achieved with standard modalities. Three different strategies have been developed to endoscopically image the small bowel, push enteroscopy, Sonde enteroscopy and interoperative enteroscopy (IOE). After a brief review of small intestinal anatomy, these techniques will be discussed in detail.

Relevant Anatomy

- The small intestine extends from the pylorus to the ileocecal valve.
- It is composed of the duodenum, jejunum and ileum.
- The average length is 67 cm, with duodenum measuring about 24 cm, the jejunum and ileum accounting for 2/5 and 3/5 respectively of the remaining length.
- The duodenum is retroperitoneal. The mesentery is lost during fetal development. The jejunum and ileum are attached to the posterior abdominal wall by a fanshaped mesentery. This short mesentery results in a coiling effect, which is one of the limiting features of small intestinal endoscopy.
- The small intestinal wall is composed of four main layers, the mucosa, submucosa, the muscular layers, and the serosa.

Indication and Contraindication

Indications

- **Occult gastrointestinal bleeding.** This is the most common indication for enteroscopy. The small intestine is the source of gastrointestinal bleeding in less than 5% of cases. However, when the bleeding is chronic and undiagnosed after standard endoscopy, the proportion of cases is larger. Patients with bleeding, presumed to be of small intestinal origin, have usually undergone an array of

Gastrointestinal Endoscopy, edited by Jacques Van Dam and Richard C. K. Wong.

diagnostic studies including upper endoscopy, colonoscopy, barium studies and occasionally arteriography and or nuclear medicine studies. Some patients will be anemic and require iron supplementation, others may require blood transfusions and even become transfusion dependent. All three types of enteroscopy may be called upon to evaluate occult bleeding.

- **Malabsorption**. Push enteroscopy may be helpful in the evaluation of diarrhea in malabsorption. The mucosa may be carefully evaluated and biopsied. Diseases that can produce malabsorption and diarrhea include celiac disease, Crohn's disease, Whipple's disease, lymphoma, eosinophilic gastroenteritis and some parasitic infections. In addition to mucosal biopsy and inspection, duodenal or jejunal aspirates can be done at the time of procedure. This may facilitate the diagnosis of parasitic infections such as giardia.
- Evaluation of **x-ray abnormalities**. An upper gastrointestinal series, small bowel follow through and or enteroclysis may reveal abnormalities in the small intestine.
- **Placement of jejunostomies**. A push enteroscope may be used to pass a jejunal feeding tube through a preexisting gastrostomy site. It may also be used in performing percutaneous endoscopic jejunostomy (PEJ placement). This is performed in a manner identical to PEG placement. It requires that a jejunal loop be close enough to the abdominal wall to allow transillumination. The placement of the endoscopic jejunotomy tubes may facilitate feeding individuals with gastroesophageal reflux disease and recurrent aspiration.
- Screening for **polyposis syndromes**. In a variety of polyposis syndromes, the small bowel may be effected by adenomatous, juvenile or hamartomatous polyps and even malignant tumors. Push enteroscopy permits careful evaluation of the mucosa to the mid jejunum. If polyps are found they may be biopsied or even removed using standard polypectomy technique. Enteroscopy is felt to be the best screening procedure for the upper gastrointestinal tract in patients with polyposis syndromes.
- **Access to the bile duct** after **Roux-en-Y** gastrojejunostomy. It is often difficult to reach the ampulla in patients who have undergone a **Roux-en-Y** gastrojejunostomy. A push enteroscope can be used to access the ampulla for evaluation and treatment of a variety of biliary and pancreatic abnormalities.

Contraindications

- As with all types of endoscopy, push enteroscopy and Sonde enteroscopy require a cooperative patient who is hemodynamically stable. There should be no evidence of severe coagulopathy. Special care should be taken in using these type of instruments in patients who have undergone esophageal or gastric surgery to avoid mucosal trauma. Specifically an overtube should be avoided in patients with an esophageal or gastric stricture.
- Intraoperative enteroscopy should only be performed in patients in whom the benefits of diagnosis and treatment of suspected small intestinal pathology outweigh the risks of laparotomy.

Equipment, Endoscopes, Devices and Accessories

- Push enteroscopy. Standard upper endoscopy permits evaluation of the small intestinal mucosa to the region between the second and third portion of the duodenum. Specialized longer instruments (enteroscopes) have been developed that may be further advanced into the small bowel. Most of the experience with small bowel endoscopy has been with pediatric or adult colonoscopy passed orally. These permit endoscopic evaluation to about 40-60 cm beyond the ligament of Treitz. Limitations of this technique include: tendency for the endoscopy to coil in the stomach limiting the depth of insertion, mucosal trauma from the wider diameter instruments that may be confused with mucosal abnormalities such as vascular malformation and patient discomfort due to the larger diameter instruments.

- Because of these problems longer, narrower instruments have been developed to permit deeper insertion into the small intestine. These enteroscopes measure 200-240 cm. Because of the tendency for these scopes to coil in the stomach, overtubes were developed. These are back loaded onto the enteroscope, and then advanced into the duodenum once the instrument is passed into the proximal duodenum. Fluoroscopy may be used to monitor passage of the overtube and to determine the depth of small bowel insertion. The push enteroscope may be advanced to the level of the mid jejunum. These instruments have an accessory channel that permits diagnostic and therapeutic options including biopsy, cautery, laser therapy and injection therapy as well as snare polypectomy. Newer video endoscopes have been developed with some increased stiffness which permits complete study without the use of an overtube.
- **Sonde enteroscopy**. In the case of chronic gastrointestinal bleeding, push enteroscopy may find the source of bleeding in 30-50% of cases. Obviously, in small bowel endoscopy the further into the small bowel one looks the greater the diagnostic yield. A Sonde enteroscope is a long, thin, flexible fiberoptic instrument which is passed either nasally or orally into the stomach and then through the small intestine by peristalsis. It has a working length of 279 cm and a tip diameter of 5 mm. It's field of view is 120°· It has a balloon at its tip that is inflated after it has passed the pylorus. Advancement through the gastrointestinal tract is passive and achieved by the traction of peristalsis. The instrument has no biopsy or therapeutic potential and there is no tip deflection. This limits the diagnostic and therapeutic capabilities of this technique.
- **Intraoperative Enteroscopy (IOE)**. IOE is the most effective technique for evaluating small bowel bleeding. However, it requires a laparotomy and should only be performed in clinical situations where the benefit of the test clearly outweighs the risk. Usually IOE is done for chronic transfusion dependent gastrointestinal bleeding of presumed small bowel etiology. However, it may also be done for acute small intestinal bleeding as well. Most of the experience with IOE has been with adult or pediatric colonoscopes passed orally at the time of laparotomy. More recently the push enteroscopes have been used for this purpose.
- The instrument is passed into the duodenum and gradually advanced through the mouth with the surgeon "pleating" the small intestines onto the instrument. The surgeon and endoscopist watch the video monitor to visualize the mucosa and the surgeon also inspects the serosal surface, which is transilluminated during the procedure. This technique permits endoscopic biopsy as well as therapy of focal lesions with cautery or laser. It also allows for definitive surgical treatment such as resection or oversew of any abnormalities found in the small intestine.
- Because of reports of mucosal trauma, perforation and even death from using standard push type scopes and colonoscopes during intraoperative enteroscopy, the technique of using a Sonde enteroscope during a laparotomy was developed. This usually permits complete evaluation of the entire small bowel with

less trauma to the small intestines. This is due to its smaller radius of curvature compared with standard and push type enteroscopes.

Technique

- **Push enteroscopy.** As for all endoscopic procedures, patients should be fasting for at least 6 h for solid foods and 24 h for liquids. In the endoscopy suite patients are placed in the left later decubitus position. Conscious sedation is given, usually with midazolam and merperidine. The push enteroscope is passed orally into the stomach and advanced into the duodenum in a manner identical to that with standard upper endoscopy. If an overtube is to be used, it is back loaded onto the instrument. The scope is passed into the duodenum, and any intragastric loop is reduced. Using generous lubrication on the endoscope, the overtube is gently passed over the scope through the pylorus into the bulb of the duodenum. The push enteroscope can then be further advanced into the small bowel by slow and gradual advancement and withdrawal similar to the technique for colonoscopy. Glucagon may be used to decrease peristalsis. Occasionally abdominal pressure or position changes may be used to facilitate passage of the instrument further into the small intestines. The mucosa is evaluated both on insertion and during removal. Special care should be taken, prior to passage of the overtube, to be sure pathology in the esophagus, stomach and proximal duodenum has not been missed at a previous endoscopy. This has been reported to occur in up to 30% of cases. Cases that may be missed at previous endoscopy include watermelon stomach, giant hiatus hernias [Cameron's syndrome], celiac disease, peptic ulcers and Dieulafoy's lesions. After the overtube is withdrawn it is common to see minor mucosal trauma. This can easily be mistaken for intrinsic bowel pathology. Hence, very careful evaluation prior to the use of the overtube is essential.

  A diagnostic push enteroscopy can usually be performed in less than 30 min. No specialized nursing needs are required. Some units use fluoroscopy for passage of the overtube and to determine the depth of insertion. The push enteroscope may be advanced 80-120 cm beyond the ligament of Treitz to the region of the mid jejunum. The learning curve for this procedure is short, and most endoscopists will be able to master this technique. The accessory channel on the push enteroscopy is available for biopsy, injection therapy, cautery, laser and polypectomy.
- **Sonde enteroscopy.** The Sonde enteroscope permits evaluation of the more distal small intestine. It can reach the ileum in 75% of cases, but the ileocecal valve is reached in only 10% of cases. Because of its lack of tip deflection and because visualization of the mucosa is only done on withdrawal only 50-70% of the mucosal surface is actually seen. The long, thin flexible instrument is passed nasally or orally into the stomach along with a standard endoscope. The gastroscope or push enteroscope is then used to "piggyback" the Sonde into the duodenum. A biopsy forceps is used to pick up a silk suture fastened to the end of the enteroscope. This allows the tip of the enteroscope to be maneuvered into the duodenum. The balloon at the tip of the Sonde instrument is inflated and the scope is then carried distally by peristalsis. The instrument is gradually advanced a few centimeters at a time. No sedation is required. Patients are placed in the supine position. The time of passage for the Sonde instrument is from 4-8 h.

The small bowel is evaluated during the withdrawal phase. Glucagon may facilitate the examination. The instrument is gradually withdrawn and air is insufflated into the lumen. The depth of insertion can be determined by fluoroscopy. Sonde enteroscopy is limited by its expense, patient discomfort, length of time for the exam to be performed and perhaps most importantly its lack of diagnostic or therapeutic capabilities. It is most useful in seeing whether or not there are focal vascular abnormalities that might respond to surgical resection or endoscopic cautery at the time of interoperative enteroscopy. If diffuse vascular lesions are found throughout the bowel surgical therapy is contraindicated. In this case other strategies such as combination therapy with estrogen and progestin, iron therapy or octreotide may be considered.

- **Intraoperative enteroscopy (IOE)**. IOE may be performed with a standard gastroscope, colonoscope, push enteroscope or as we have previously reported a Sonde enteroscope. A standard laparotomy is performed under general anesthesia. All adhesions are lysed. The surgeon carefully evaluates the stomach, duodenum, small and large intestine. This may reveal gross lesions such as tumors, Meckel's diverticulum or a large vascular malformation. In cases where a push enteroscope or pediatric colonoscope is used, the instrument is passed into the duodenum in the standard fashion. Because of the diameter of the instruments the endotracheal tube may need to be temporarily deflated to facilitate esophageal intubation. The scope is then passed into the small intestine. The surgeon then telescopes the small intestine over the scope. Careful mucosal examination is done on insertion and withdrawal. These instruments often will not reach the ileocecal valve due to their radius of curvature being greater than that of the mesentery. If the scopes are pushed there may be trauma to the mesentery and mucosa. In cases where it is not possible to reach the terminal ileum, an enterotomy can be made and a sterilized endoscope can be passed to complete the study. These instruments cause mucosal trauma, mesenteric stretching, prolonged ileum, perforation and even death. These complications plus the failure to reliably visualize the entire small bowel led to a trial using the Sonde instrument for intraoperative enteroscopy. Surgery as a sole means of detecting the source of chronic gastrointestinal bleeding is successful only 50% of the time or less. The Sonde endoscope is inserted into the stomach via a 9 mm endotracheal tube placed in the esophagus. Care is taken to avoid coiling of the instrument. The surgeon locates the tip of the scope in the stomach and manipulates it into the duodenum. The balloon at the tip of the scope is inflated in the duodenum. The surgeon then uses the balloon as a "handle" to manipulate the scope. The small intestine is then telescoped over the enteroscope. The surgeon may need to perform a Kocher maneuver of the duodenum to facilitate passage of the scope round the duodenum. As the instrument is passed through the small intestine, the mucosa of the small bowel is best visualized using the "air trapping" technique. An assistant isolates a 10-15 cm segment of intestine and pinches the distal portion thus trapping enough air to keep the lumen open and permitting careful endoscopic inspection. Intestinal clamps are not used. A 60 ml syringe instills enough air to keep the lumen open. All mucosal abnormalities are photographed and serosal sutures are used as markers. Once the instrument is removed, the appropriate definite procedure is performed. This may include oversewing vascular malformations or resection of abnormal segments of small intestine. The endoscopic portion of the procedure takes less than 30 min. This technique permits safe and thorough evaluation of the entire small intestine with a high diagnostic yield and good therapeutic efficacy. Patients tolerate this type of intraoperative enteroscopy with less abdominal discomfort and ileus.

Outcome

- **Push enteroscopy**. In the evaluation of chronic gastrointestinal bleeding of presumed small bowel etiology, push enteroscopy will reveal a source of bleeding in 30-80% of patients. It is important to keep in mind that up to 30% of abnormalities found are within the reach of the standard gastroscope. The most frequent lesions found in the small intestine are arteriovenous malformations (AVMs). Other causes of bleeding, especially in younger patients, include small bowel tumors, both malignant and benign tumors, and Meckel's diverticulum. Other lesions which have been found include NSAID ulcers and strictures, small intestinal varices, Crohn's disease, and radiation enteritis. In patients with persistent transfusion dependent GI bleeding with an unrevealing push enteroscopy, Sonde enteroscopy may be helpful.
- **Sonde enteroscopy**. The Sonde instrument may find a source of obscure GI bleeding in up to 40% of cases. The most common lesions seen are arteriovenous malformations. However, the significant limitation of this test is that no therapy is possible. In addition, it is not possible to even mark the bowels at a site of an abnormality for future identification at laparotomy and intraoperative enteroscopy.
- **Intraoperative enteroscopy**. In the evaluation of chronic gastrointestinal bleeding, intraoperative enteroscopy has the highest yield of all diagnostic tests. Rates of between 70% and 90% have been reported. The most common findings are vascular malformation. Other abnormalities which can be found include small bowel ulcerations, radiation enteritis and neoplasia. The vascular malformation may either be oversewn or resected. Endoscopic therapy is also possible. In our series on intraoperative enteroscopy, long-term follow-up has revealed excellent control of small intestinal bleeding.

Complications

- Enteroscopy has a slightly higher risk of complications compared with standard endoscopy. The risks which are similar to standard endoscopy include medication reaction, discomfort, bleeding and perforation. The risks of push enteroscopy are slightly higher primarily due to the use of an overtube. Overtubes have been associated with mucosal trauma and esophageal and pharyngeal perforations. As new enteroscopes have been developed, we use the overtube less frequently and find a high degree of safety and depth of insertion. The Sonde enteroscope primarily causes patient discomfort during the time of the study. The major problem is that the study takes several hours. However, most patients tolerate the Sonde enteroscopy quite well. In intraoperative enteroscopy the risks of the procedure include those of a standard laparotomy with general anesthesia. In addition, using a pediatric colonoscope or push enteroscope has been associated with mucosal trauma, prolonged ileus, perforation or even death. Utilizing the Sonde enteroscope for intraoperative enteroscopy, we have found that patients generally tolerate the procedure well.

Summary

The small intestine is an important source of pathology, particularly in chronic gastrointestinal bleeding and malabsorption. It may be necessary to access the small intestine to place jejunostomies. In addition, in patients who are status post Roux en Y gastrojejunostomy, push enteroscopy may permit manipulation of the pancreatic and biliary tree. Screening for small intestinal polyps and the polyposis syndromes can be effectively carried out with push enteroscopy. Sonde enteroscopy may be useful in evaluating the small intestine in patients with chronic gastrointestinal bleeding. However, its limitations may outweigh its usefulness. In our opinion intraoperative enteroscopy is the procedure of choice in patients with transfusion-dependent chronic occult gastrointestinal bleeding of presumed small intestinal etiology. We believe that a Sonde enteroscope is less traumatic than a standard colonoscope or push enteroscope. It will permit evaluation of the small intestine with a decrease in the complication rate. Cooperation between the surgeon and the endoscopist is essential. In carefully selected patients we believe this to be an excellent modality in the diagnosis and management of chronic gastrointestinal bleeding.

Selected References

1. Waye JD. Enteroscopy. Gastrointest Endosc 1997; 46:247-256.
   This is an excellent review of all three types of enteroscopy. The data is carefully reviewed. There is an excellent discussion by that author.
2. Rutgeerts T, Wilmer A. Push enteroscopy: Technique, depth and yield of insertion. In: Barkin J, ed. Gastrointest Endosc Clinics North Am 1996:759-776.
   This is a thorough and excellent review of the technique of push enteroscopy.
3. Lewis B. Small intestinal bleeding. In: Gastroenter Clinics North Am 1994:67.
   This is an excellent overview of evaluation and management of small intestinal bleeding in general. Different diagnostic and therapeutic options are discussed. In addition, the different etiologies for small intestinal bleeding are reviewed.
4. Lopez M, Cooley J, Petros J et al. Complete intraoperative smallbowel endoscopy in the evaluation of occult gastrointestinal bleeding using the sonde enteroscope. Arch Surg 1996; 131:272-276.
   This is a review of our experience using intraoperative enteroscopy in patients with chronic gastrointestinal bleeding. The technique is well outlined.
5. Hoffman J, Cave DR, Birkett D. Intraoperative enteroscopy with a sonde intestinal fiberscope. Gastrointest Endosc 1994; 40:229-230.
   This is the initial report of the use of Sonde enteroscopy for intraoperative enteroscopy.
6. Gostout C. Sonde enteroscopy: Technique, depth of insertion and yield of lesions. In: Ed Gastrointest Endosc Clinics North Am:777-792.
   This is an in-depth review of Sonde enteroscopy.
7. Cave D, Cooley J. Intraoperative enteroscopy: Indications and techniques. In Ed: Gastrointest Endosc Clinics North Am 1996:793-802.

CHAPTER 17

Flexible Sigmoidoscopy

Richard C.K. Wong and Jacques Van Dam

Introduction

There are few invasive tests that can be completed in under 10 minutes, be performed by both digestive disease specialists, trained primary care physicians (including some nurses), and which offer the ability to reduce mortality from colorectal cancer. Flexible sigmoidoscopy is one such test. Not only is this technique central to current guidelines of screening for colorectal neoplasia, it is also important in the elective and occasionally in the emergency setting for the detection of distal colonic disease. In general however, flexible sigmoidoscopy should not be performed for indications that warrant colonoscopy (e.g., unexplained lower gastrointestinal bleeding, iron deficiency anemia or positive fecal occult blood test in individuals over the age of 50). Flexible sigmoidoscopy is primarily a diagnostic test with the ability to sample tissue by performing endoscopic biopsy. In some cases however, digestive disease specialists can use the flexible sigmoidoscope as a therapeutic tool, for instance, in band ligation of internal hemorrhoids and in the emergency decompression of sigmoid volvulus.

Relevant Anatomy

- dentate line, located approximately 2 cm proximal to the anal verge, marks the transition between columnar epithelium proximally (innervated by the autonomic nervous system) and squamous epithelium distally (innervated by the somatic nervous system) The patient can experience painful somatic sensation distal to the dentate line
- rectum, located proximal to the anal canal, contains three semicircular valves (Valves of Houston)
- rectosigmoid junction is a sharp bend which begins approximately 15 cm proximal to the anal verge
- sigmoid colon, located proximal to the rectosigmoid junction, approximately 40 cm in length in adults in Western countries.
- descending colon, is a relatively straight tubular portion of the colon approximately 20 cm in length. It connects the sigmoid colon to the splenic flexure.
- splenic flexure, where the colon takes a sharp bend to form the transverse colon (characterized by triangular-appearing folds). The splenic flexure is reached in approximately 22% of cases at full insertion using a 60 cm long, 16 mm diameter flexible sigmoidoscope. The vascular supply to the colon includes the so-called "watershed" areas (splenic flexure and rectosigmoid junction) which are particularly susceptible to ischemic injury.

Gastrointestinal Endoscopy, edited by Jacques Van Dam and Richard C. K. Wong.

Indications

Screening for Colorectal Cancer

- Adenocarcinoma of the colon and rectum is the second most common fatal cancer in the United States. It has an incidence of approximately 140,000 cases per year and a mortality of approximately 55,000. Early detection and treatment are essential in reducing this incidence and mortality of the disease.[1,2]
- Endorsed by U.S. Preventative Services Task Force, American Cancer Society, National Cancer Institute
- Recommended for men and women over the age of 50 every 5 years until the time when screening for colorectal cancer will not benefit the patient
- The 60 cm rather than the 35 cm flexible sigmoidoscope is recommended
- Many authorities recommend biopsy of polyps under 1 cm : colonoscopy should then be offered if adenomas or cancers are found. Patients with polyps greater than 1 cm should be offered colonoscopy without the need for biopsy.
- Adenomatous polyps (found in half to two thirds of patients) have malignant potential whereas hyperplastic polyps (found in 10 to 30% of patients) have no clinical significance[3]

"Elective" Flexible Sigmoidoscopy

- Test of choice in patients suspected of having distal colonic disease or anorectal disorders, e.g., young patient with history of hemorrhoidal bleeding
- Can be used in selected patients with severe bleeding from internal hemorrhoids as a therapeutic tool to perform endoscopic band ligation
- Has an important role in evaluating some patients with persistent diarrhea
- Can diagnose pseudomembranous colitis or melanosis coli
- With endoscopic biopsies can diagnose microscopic colitis, amyloidosis, graft-versus-host disease or idiopathic inflammatory bowel disease

"Emergency" Flexible Sigmoidoscopy

- Can be used for emergent endoscopic decompression of a sigmoid volvulus and for the diagnosis of ischemic colitis and pseudomembranous colitis

Contraindications

- Flexible sigmoidoscopy should never be performed in patients with acute peritonitis or when the risk of bowel perforation is high (Table 17.1)
- The decision to perform the examination must be individualized, and the diagnostic gain/benefit must be weighed against the risk of the procedure

Endoscope

- Short, 30-35 cm
- Intermediate, 45 cm
- Long, 60-77 cm (recommended as it can detect significantly more pathology than the shorter lengths)

Preparing the Patient

- History and physical examination should be performed prior to the procedure
- Indications, alternatives, benefits and risks of the procedure should be reviewed with the patient and informed consent obtained

Table 17.1. Contraindications to flexible sigmoidoscopy

Absolute	
Acute peritonitis	Perforated viscus
Acute, severe diverticulitis	Severely uncooperative patient
Unconsenting, fully competent patient	
Relative	
Bowel infarction	Fulminant colitis
Toxic megacolon	Inadequate bowel preparation
Poor cardiopulmonary reserve	Recent major abdominal surgery
Recent myocardial infarction	Late stages of pregnancy
Serious cardiac arrhythmia	Large abdominal aortic aneurysm

- The physician should explain, in a step-by-step manner, what the examination involves and what the patient may feel during the procedure. In particular, the patient should be warned that there may be discomfort from air insufflation and be ready to inform the endoscopist if there is severe pain.

Bowel Preparation

- There are several types, the phosphate enema regimen (Table 17.2) produces satisfactory results
- On rare occasions, a more extensive preparation may be needed which may include a 10 fl oz bottle of magnesium citrate prior to the examination
- The phosphate-based enemas should be avoided in individuals with significant renal insufficiency and in children. A bowel preparation may not be needed in patients with severe diarrhea.

Antibiotic Prophylaxis

- In general, antibiotic prophylaxis is not required for diagnostic flexible sigmoidoscopy with or without endoscopic mucosal biopsy[4]
- In certain "high risk" individuals (those with prosthetic heart valves, prior history of endocarditis, surgically constructed systemic-pulmonary shunts and synthetic vascular grafts less than one year old) there is insufficient data to make firm recommendations and the endoscopist may decide on a case-by-case basis.

Sedation

- Not required for flexible sigmoidoscopy in the vast majority of cases

Positioning the Patient and the Endoscopist

- Sigmoidoscopy is best performed with the patient in the Sims or left lateral decubitus position. The hips and knees are partially flexed and the right knee is positioned above the left one. The endoscopist should stand between the light source and the back of the patient.

Universal Blood/Body Fluid Precautions

- The endoscopist and any assistants should adhere to strict universal blood/body fluid precautions. Protective gowns, protective eyewear, gloves and mouth guards should be worn at all times.

Table 17.2. Bowel preparation for flexible sigmoidoscopy

- Two 4.5 fl oz phosphate enemas* (e.g.,Fleet's) given within two hours of the exam
- NPO after the enemas except for medication
- Unprepped, if patient has severe diarrhea

* avoid phosphate-based enemas in individuals with significant renal insufficiency or in children.

Technique

- A thorough examination can be performed in under 10 minutes using the one-person technique
- The trainee will initially take much longer and may wish to use the two-person technique initially. On average, trainees require a minimum of 25 supervised procedures before being competent in performing diagnostic flexible sigmoidoscopy[5]
- Check the equipment
- The endoscopic image should be assessed and focused, the air/water and suction checked and both lock dials should be placed in the unlocked position
- Examination of the perianal area: important to rule out perianal pathology such as viral condylomata, hemorrhoids, fissures and fistulas
- Digital rectal examination performed using a well-lubricated gloved finger. This serves to lubricate the anal canal, relax the rectal sphincter, detect pathology in the region (including the prostate), and give an initial assessment of the effectiveness of the bowel preparation.
- Intubation of the rectum: the lubricated tip of the scope is gently introduced into the rectum by flexion of the right index finger, guiding it into the anus at a 90° angle. This is better tolerated than an endon 180° approach.

Visualizing the Rectum

- Once the tip of the scope is in the rectum, a so-called mucosal "redout" is usually seen. This is caused by the tip of the scope being situated on the rectal mucosa. Vision can be restored by slight withdrawal of the scope and by air insufflation. Table 17.3 lists some visual clues in finding the lumen of the bowel.
- The experienced endoscopist utilizes the *least* amount of air insufflation during the examination. Too much air insufflation can cause undue patient discomfort and can make the procedure more difficult by lengthening bowel loops.
- Most of the evaluation of the colorectal mucosa is performed during *withdrawal* of the instrument, as in colonoscopy. Liquid in the rectum should be aspirated via the sigmoidoscope for a clearer view.
- Aspiration of solid or semisolid material should not be attempted as this will obstruct the channel. Temporary misting of the lens can be remedied by depressing the water button followed by a short burst of suction.
- Normal rectal mucosa appears glistening and pink in color with a fine branching vascular pattern. Occasionally, the use of enemas in the bowel preparation can give rise to artefactual changes in the rectal mucosa that appear as inflammation. Thus, a patient with diarrhea and suspected idiopathic inflammatory bowel disease is better examined in the unprepped state.

Table 17.3. Clues to finding the bowel lumen

- The lumen will be in the direction of the darker mucosa (i.e., the point furthest from the tip illumination)
- If a curvilinear mucosal fold is seen, the lumen will be found behind it
- In a segment of colon with evident concave arcs representing haustral folds, the lumen will be in the center of an imaginary circle formed by these arcs
- In most instances, the lumen will be found away from the mucosal "redout"
- If multiple diverticula are present, the true lumen will be found in the presence of haustral folds

Advancing Beyond the Rectosigmoid Junction

- After advancing the endoscope past the three rectal valves of Houston, a sharp bend is encountered at approximately 15 cm from the anal verge. This is the rectosigmoid junction and represents the area of most difficulty during the sigmoidoscopic examination. With practice, the following maneuver will be found most useful. First, advance the endoscope beyond the last valve of Houston, then deflect the tip upwards and, with gentle clockwise torquing, slowly advance the endoscope beyond the rectosigmoid junction.

The Sigmoid Colon

- Can be another difficult area to negotiate as it is made up of a series of semilunar valves and the lumen will be partially hidden behind each valve
- Advance slowly, always keeping the lumen in view, with a combination of gentle torquing and forward motion (with the right hand on the scope) and manipulation of the up/down deflection (with the left thumb)
- Short bursts of air can be useful. A mucosal "redout", indicates that the tip of the endoscope abuts the mucosa. Pull back and reassess where the lumen is. If a mucosal "whiteout" is seen, the examiner *must* pull the scope back a little as it means that undue pressure is being applied to the mucosa causing it to blanch. Further forward pressure at this point could result in perforation.

The Descending Colon/Splenic Flexure

- The descending colon resembles a long, fairly straight tube and the splenic flexure has a typical bluish hue to its color. With a 60 cm long and 16 mm diameter scope, the splenic flexure is reached in about 22% of the time with full insertion.

Withdrawal of the Scope

- Most important part of the procedure as the most detailed views of the mucosa are obtained during withdrawal
- Inspection of the full 360° of the lumen and behind all folds. Sometimes it may be necessary to advance the endoscope again and then pull back if a corner was poorly visualized.

Retroflexion in the Rectum

- Important to perform this maneuver as small lesions in the distal rectum, near the anal canal, may be easily missed on forward view
- After informing the patient that he/she may feel transient rectal discomfort, the endoscope is positioned at approximately 10 cm from the anal verge, the right/

Table 17.4. "Pearls" in reducing procedurerelated complications

- Careful patient selection
- Close trainee supervision
- Always know where the lumen is before advancing the scope
  " If in doubt, pull it out !"
- Minimize air insufflation
- Pull back if patient complains of severe discomfort
- Withdraw scope on mucosal "whiteout"
- Advance very cautiously in the presence of diverticulosis
- Know when to stop the exam and/or ask for help

left control is deflected maximally counterclockwise, locked, and the instrument is very carefully advanced as the tip is slowly deflected upwards. With some air insufflation and rotational torque, the anorectal junction should clearly be seen around its entire circumference. Once this has been completed, the endoscope should be withdrawn by returning the dials to the neutral position and the shaft slowly pulled back as air is being suctioned from the lumen.

- Endoscopic biopsy should be performed only by an experienced operator. Note that most cases of perforation or bleeding occur after a biopsy only biopsies without electrocautery should be performed so-called "hot biopsies" utilize electric current and may lead to an intraluminal gaseous explosion: such biopsies should be performed only during colonoscopy when the colon has been thoroughly cleansed of combustible gases

Complications

- Major complications are rare, but include vagally-mediated reactions, hemorrhage, infection and perforation
- The perforation and hemorrhage rates are approximately 1 in 8,000 and 1 in 17,000 respectively
- Most complications occur during the early training period of the endoscopist and rapidly decline with experience (Table 17.4)

Selected References

1. Selby JV, Friedman GD, Quesenberry CP Jr et al. A case-control study of screening sigmoidoscopy and mortality from colorectal cancer. N Engl J Med 1992; 326:653-657.
2. Newcomb PA, Norfleet RG, Storer BE et al. Screening sigmoidoscopy and colorectal cancer mortality. J Natl Cancer Inst 1992; 84:1572-1575.
   References 1 and 2: Two case-control studies showing that patients with fatal colorectal cancer were less likely to have had screening sigmoidoscopy than comparable control groups.
3. Winawer SJ, Fletcher RH, Miller L et al. Colorectal cancer screening: Clinical guidelines and rationale. Gastroenterology 1997; 112:594-642.
   Comprehensive review of colorectal cancer screening strategies and their rationale.
4. Antibiotic prophylaxis for gastrointestinal endoscopy. Gastrointest Endosc 1995; 42:630-635.
   Guidelines on the necessity of antibiotic prophylaxis for gastrointestinal endoscopy.
5. Principles of training in gastrointestinal endoscopy. Gastrointest Endosc 1992; 38:743-746.
   American Society for Gastrointestinal Endoscopy (ASGE) guidelines regarding threshold for assessing competence in performing gastrointestinal endoscopy.

CHAPTER 18

Colonoscopy

Douglas K. Rex

Introduction

Colonoscopy is the most powerful visceral cancer prevention technique in use in clinical medicine. In gastroenterologists' practice it is the most common endoscopic procedure performed. Colonoscopy should be performed by individuals who have completed training in a gastrointestinal fellowship or surgical residency or an equivalent. The American Society for Gastrointestinal Endoscopy recommends that at least 100 colonoscopies be performed under supervision before trainees are assessed for competence. A recent study suggested that 200 colonoscopies is necessary before adequate technical skills and ability to recognize pathology are achieved. Individuals performing colonoscopy independently should be able to achieve cecal intubation in > 90 % of patients and should be able to perform polypectomy and biopsy.

Practical Colonoscopic (Endoscopic) Anatomy

- Cecum. Intubation of the cecum can be verified with virtual certainty by identification of the appendiceal orifice and the ileocecal valve and its orifice (Fig. 18.1). Additional landmarks of value are the cecal sling fold, formed by the lateral taenia coli crossing the cecal pouch, the "crow's foot" formed by the confluence of the taenia around the appendiceal orifice, and intubation of the terminal ileum. Transillumination of the right lower quadrant or right groin is complimentary but unreliable.
- Right colon. The right colon has a thin wall which increases the risk of perforation with therapeutic procedures. The luminal caliber is often large.
- Transverse colon. The longitudinal muscle of the colon is grouped into three bundles called taenia coli. These give the lumen a triangular appearance, particularly in the transverse colon. However, the use of anatomic landmarks to localize a colonic lesion is reliable only when the cecum has been verified by landmarks and is still in view, or in the distal 25-30 cm of the colorectum, which includes the rectum and the distal sigmoid.
- Left colon anatomy (including rectum, sigmoid, descending colon). See chapter on flexible sigmoidoscopy.

Indications

- Suspected colonic bleeding
- Positive fecal blood test
- Nonemergent hematochezia
- Acute lower gastrointestinal hemorrhage
- Iron deficiency anemia of unknown cause

Gastrointestinal Endoscopy, edited by Jacques Van Dam and Richard C. K. Wong.

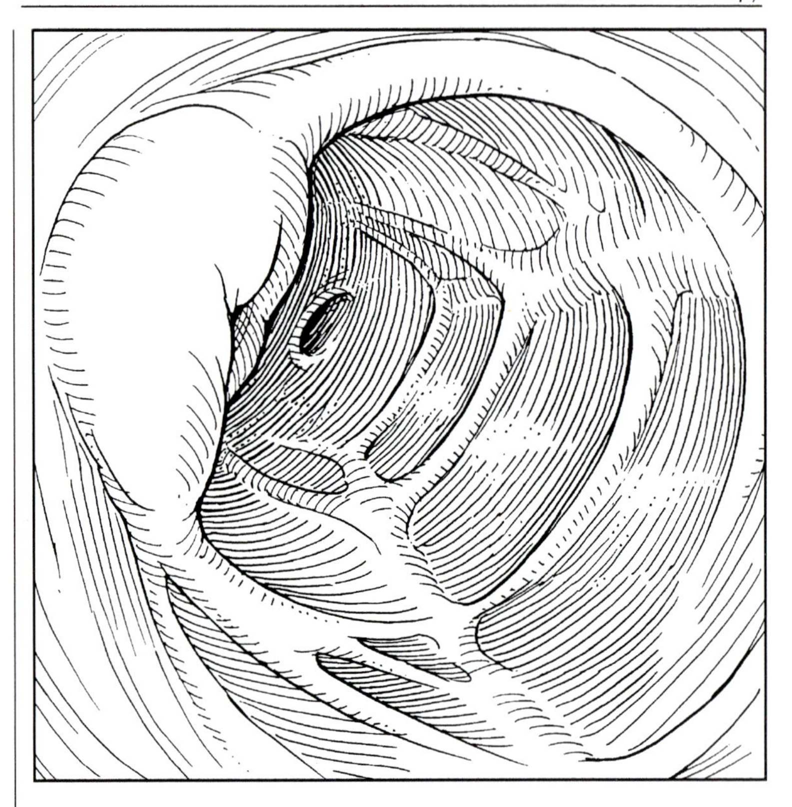

Figure 18.1. Endoscopic appearance of the cecum.

- Melena following a negative esophagogastroduodenoscopy
- Evaluation of an abnormal barium enema showing a polyp, mass, stricture, ulcer or suspected inflammatory process.
- Following a screening sigmoidoscopy demonstrating an adenomatous polyp
- Inflammatory bowel disease
- To clarify the diagnosis or to determine the extent or activity of disease
- For cancer surveillance in ulcerative colitis extending proximal to the splenic flexure, beginning after 8-10 years of symptoms, at an interval of 1-3 years until 20 years, then yearly or more frequently as determined by the presence of other risk factors (primary sclerosing cholangitis or a family history of colorectal cancer).
- For cancer surveillance in left-sided ulcerative colitis, after 15 years of symptoms, at intervals as described for pancolitis
- In surveillance for cancer in Crohn's disease
- Chronic watery diarrhea of unknown etiology
- Surveillance after resection of adenomatous polyps
- At three years for most patients after clearing colonoscopy

- At five years if clearing colonoscopy shows only 1 or 2 tubular adenomas, and the patient has a negative family history for colorectal cancer
- After removal of large sessile adenomas, at 2-6 months intervals until complete resection is verified, and then in 1 year to rule out a late recurrence
- At one or two years after clearing colonoscopy in which "numerous" adenomas are removed
- Surveillance after surgical resection of colorectal cancer
- Clearing colonoscopy is indicated preoperatively in patients with nonobstructing cancer, at 2-3 months after surgery in a patient with an obstructing cancer even if a preoperative barium enema was negative
- Screening for colorectal cancer
- At 3-5 year intervals beginning at age 40 or ten years younger than the youngest affected relative in persons with multiple first-degree relatives with cancer and/or adenomas or a single first-degree relative diagnosed at age < 60 years with cancer or adenoma
- In persons with suspected hereditary nonpolyposis colorectal cancer (HNPCC) or an HNPCC-like history, at 2 year intervals beginning at age 20 until age 40, and then annually
- Every 5 years beginning at age 40 in persons with a single first-degree relative diagnosed with cancer or adenoma after age 60
- In average-risk persons, at ten year intervals beginning at age 50
- Intraoperative
- To locate a polypectomy site, a polyp or tumor for resection (the need for this is obviated by India ink labeling during a preoperative colonoscopy)
- To evaluate an anastomosis for leaks
- To identify a bleeding lesion
- Endosonography
- Balloon dilation of colonic strictures
- Placement of colonic stents
- Decompression of acute colonic pseudo obstruction (Ogilvie's syndrome)
- Foreign body removal

Contraindications

- Absolute contraindications to diagnostic or therapeutic colonoscopy
  - Documented or suspected perforation
  - Fulminant colitis
  - Patient's refusal to give consent
- Absolute contraindications to therapeutic procedures including polypectomy, dilation, and destruction of vascular lesions
  - Severe thrombocytopenia
  - Systemic anticoagulation. (Note that aspirin and other nonsteroidal anti-inflammatory drugs are not a contraindication to polypectomy or other therapeutic procedures.) Mucosal biopsy can be performed in the setting of therapeutic anticoagulation.
- Relative contraindications to diagnostic or therapeutic colonoscopy
  - Recent myocardial infarction
  - Unstable cardiopulmonary disease
  - Hemodynamic instability; prior to attempted resuscitation
  - Inability to obtain informed consent

- Uncooperative patient
- Severe granulocytopenia
- Thrombocytopenia (< 50,000 for therapeutic colonoscopy; 30,000 for diagnostic colonoscopy)
- Severe splenomegaly
- Large abdominal aortic aneurysm
- Late stage pregnancy

Equipment

Most colonoscopes in use in the United States are produced by the three Japanese manufacturers, Olympus, Pentax and Fujinon.

- Colonoscope imaging
  - Imaging in colonoscope systems is accomplished by one of two fundamental methods: fiberoptic or video (Fig. 18.2).
  - In a fiberoptic system light reflected from the colon walls is focused by a lens system onto a fiberoptic bundle which transmits the image to another lens system in an eyepiece. The fiberoptic image is viewed directly by the colonoscopist in the eyepiece.
  - In a video imaging system reflected light is focused by a lens system onto a CCD (charged couple device – this is a television chip) (television chip). Each light stimulated pixel in the CCD chip transmits an electronic signal via wires to a video processor which converts the spatially oriented pixel signals into a signal capable of display on a video monitor.
  - Video systems hold distinct advantages over fiberoptic systems for teaching, performance of therapeutic procedures by the assistant, image recording and management, report generation, and data management. However, they are considerably more expensive than fiberoptic systems.
- Colonoscope illumination
  - Colonic illumination in both fiberoptic and video systems is achieved by light transmission from a bulb in the light source through the instrument via a fiberoptic bundle.
  - Olympus systems illuminate the colon with white light and utilize a corresponding CCD color chip.
  - Pentax and Fujinon systems utilize a rotating color filter wheel in the path of white light from the light source to send alternating red, green and blue light down the illumination system. A black and white CCD chip transmits sequential information on the reflected light which is sent to the video processor where it is used to reconstruct a color image. Each system produces acceptable images for clinical use.
- Colonoscope directional control
  - All commercially available colonoscopes utilize rotating wheels on the control head to manipulate wires in the insertion and bending portions that deflect the bending section 180∞ in both the up and down directions and 160∞ in the right and left directions. When an instrument is deflected in both the maximum up or down direction, and the maximum right or left direction, it will hairpin on itself (Fig. 18.3).
- Colonoscope air, suction, instrument access
  - Air insufflation in commercial endoscopes is sufficiently strong to insufflate air beyond cecal bursting pressure. Despite this, actual perforation from air

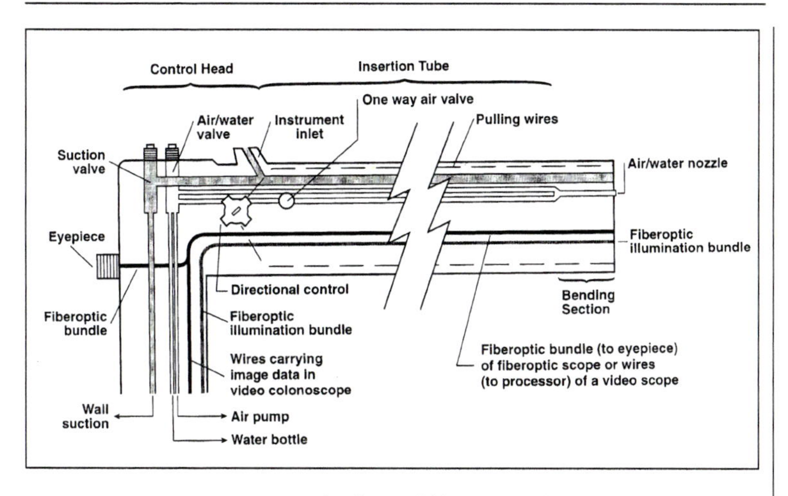

Figure 18.2. Basic components of video and fiberoptic colonoscopes.

insufflation is rare. Nevertheless discretion in insufflating air is appropriate and will improve patient tolerance.

- Olympus systems have variable air insufflation controls, while Pentax systems generally insufflate at one high rate.
- All systems allow washing of the lens systems on the colonoscope tip by full depression of the air/water valve on the control head.
- Pentax and Fujinon have accessory "water jet" systems which allow foot pedal controlled spraying of water on the colonic mucosa to remove adherent debris or blood.
- The diameter of the instrument channel determines the efficiency of suctioning and the size of accessories that can be passed the instrument. Therapeutic channels are generally at least 3.7 mm in diameter and allow easy passage of 3.2 mm (10 F) diameter instruments. Pentax makes a two channel therapeutic instrument which can facilitate performance of therapeutic colonoscopy in the setting of active bleeding.

Accessories

- **Forceps**
  - Biopsy forceps may be disposable or reusable. Whether reusable forceps can be adequately disinfected is questionable, and therefore disposable accessories are preferred.
  - Variations in forceps design include needles, serrated jaws, and fenestrated cups (Fig. 18.4). Needles or spikes improve localization of forceps placement, serrations decrease the tearing action when the forceps pull the mucosa off, and fenestration increases the size of the specimen obtained.
  - Commercially available forceps sample the mucosa and occasionally submucosal tissue. Muscularis propria is not obtained.
  - Large capacity or "jumbo" forceps are preferred for biopsies in surveillance ulcerative colitis, as they substantially provide large specimens. Therapeutic colonoscopes are needed to pass jumbo forceps.

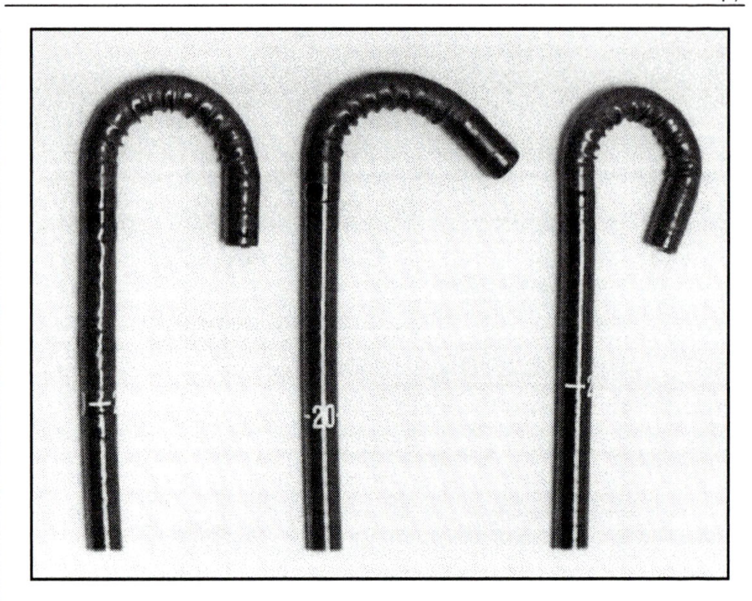

Figure 18.3. Tip defection is 180° up and down in the colonoscope (left, 160° right and left (center) and maximum deflection up or down and right or left will hairpin the bending section (right).

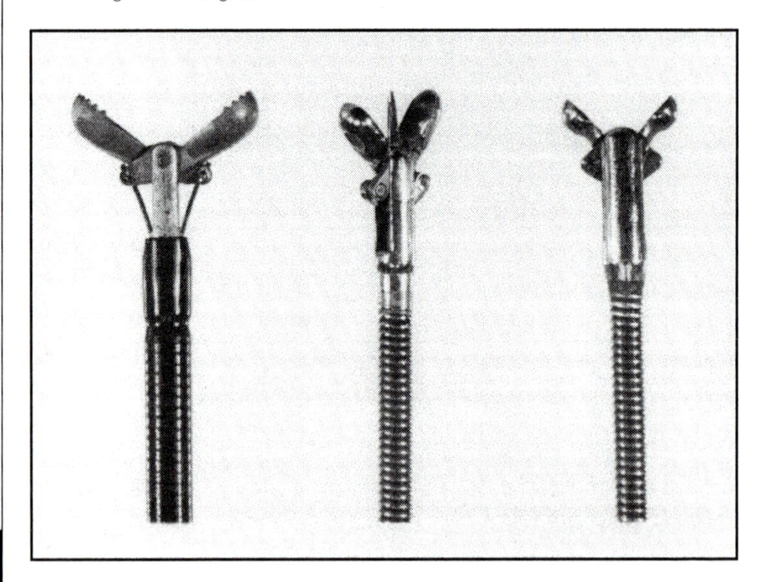

Figure 18.4. Variations in forceps design. Serrated jaws (left) and outer nonconductive covering in a hot forceps. Needled forceps with fenestrations (center) and standard forceps (right).

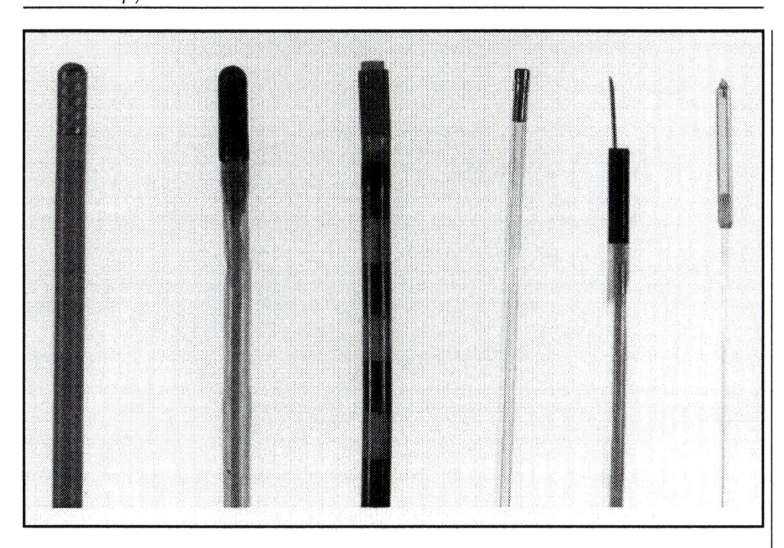

Figure 18.5. Instruments of the therapeutic colonoscopist. From left to right. Multipolar forceps, heater probe, argon plasma coagulation fiber, Nd:YAG fiber, injection needle, photodynamic therapy fiber.

- **Hot Forceps** are designed to take biopsy specimens and deliver ablative electrocautery. Most forceps are monopolar, but bipolar hot forceps are available.
- **Snares**
    - Polypectomy snares may be monopolar or bipolar.
    - Bipolar snares have been only recently introduced and may have inferior performance characteristics in removal of large polyps.
    - Nearly the entire market in polypectomy snares in the United States is for monopolar instruments.
    - Polypectomy snares differ by size (diminutive, standard, jumbo, etc.). Diminutive snares are preferred for routine use as approximately 90% of all colon polyps removed are < 1 cm in size. Diminutive snares are sufficiently easy to use that in many experts hands they have replaced hot forceps.
    - The shape of snares may be oval, crescentic, or hexagonal. Most colonoscopists use oval snares.
    - Removal of flat sessile polyps can be facilitated by barbed or spiked snares, which contain small barbs that dig into the mucosa and keep the snare from sliding over a flat neoplasm.
- **Injection needles**
    - Injection needles (Fig. 18.5) are used for submucosal saline injection during polypectomy (see below), tattooing of large polyps and cancers (see below), and for injection of dilute epinephrine and saline during active colonic hemorrhage. A syringe containing the material to be injected is connected to the catheter and the injected material flows through a plastic sheath and then into the colon wall through an endoscopically positioned 23 or 25 gauge needle.

- Multipolar cautery probes
  - The BICAP (American-McGaw) and GOLD (Microvasive) probes are **multipolar electrocautery** instruments. The tip of the probe contains electrodes for both delivery and receipt of current (Fig. 18.6). These devices are highly useful contact electrocautery probes. They deliver excellent electrocautery for hemostasis as well as lesion ablation and are relatively safe because of minimal depth of injury.
- Heater probe. The **heater probe** has a Teflon coated metal tip with a heating element inside the tip that produces coagulation necrosis by direct heating of tissue (Fig. 18.6). The uses of the instrument are similar to the multipolar cautery probes.
- Nd:Yag Laser. The lasing medium neodynium: yttrium aluminum, garnet, generates a powerful laser light with relatively high depth of penetration. **Nd:YAG laser** is appropriate for palliation of bleeding and obstruction in inoperable colorectal cancer or patients who refuse surgery, for destruction of residual flat benign adenomas in very large sessile polyps that cannot be removed by piecemeal polypectomy (see below), and for destruction of vascular lesions in the rectum and sigmoid colon. The laser light is delivered through a special endoscopic fiber which may be either noncontact (Fig. 18.6) or contact (sapphire tipped). Usually noncontact probes are used.
- **Argon plasma coagulation**. The argon plasma coagulator (APC) is an electrocautery device. Inert argon gas is passed down the endoscopic probe. The inert gas is an excellent electrical conductor and an electrical spark at the tip of the probe causes current to jump through the gas to the nearest available tissue.
  - The instrument is generally used in a noncontact mode.
  - Though formal studies are limited at this time, APC appears useful for destruction of vascular lesions and destruction of residual flat adenoma after piecemeal polypectomy.
  - APC is the only endoscopic ablation instrument that can deliver noncontact electrocautery in either en face or and tangential application.
  - APC has less depth of penetration than Nd:YAG laser, and therefore it appears less preferable for cancer ablation.
- Self-expanding metal stents. Patients with inoperable colorectal cancer with impending obstruction may be palliated by placement of **self-expanding metal stents.**.These devices are placed over guidewires positioned using endoscopic and fluoroscopic control.

Complications

- Perforation
  - Mechanical disruption of the colon from instrument passage generally occurs at the rectosigmoid junction or in the sigmoid colon. The expected perforation rate is 1/3,000 to 1/5,000 diagnostic colonoscopies.
  - The hole in the bowel created by mechanical disruption is often large and should generally be managed surgically, particularly if the peritoneal cavity is identified during withdrawal of the colonoscope or the perforation is recognized early, or the bowel preparation is poor.
  - Perforation may also occur from therapeutic colonoscopy, and may be related to polypectomy, stricture dilation, or destruction of vascular malformations.

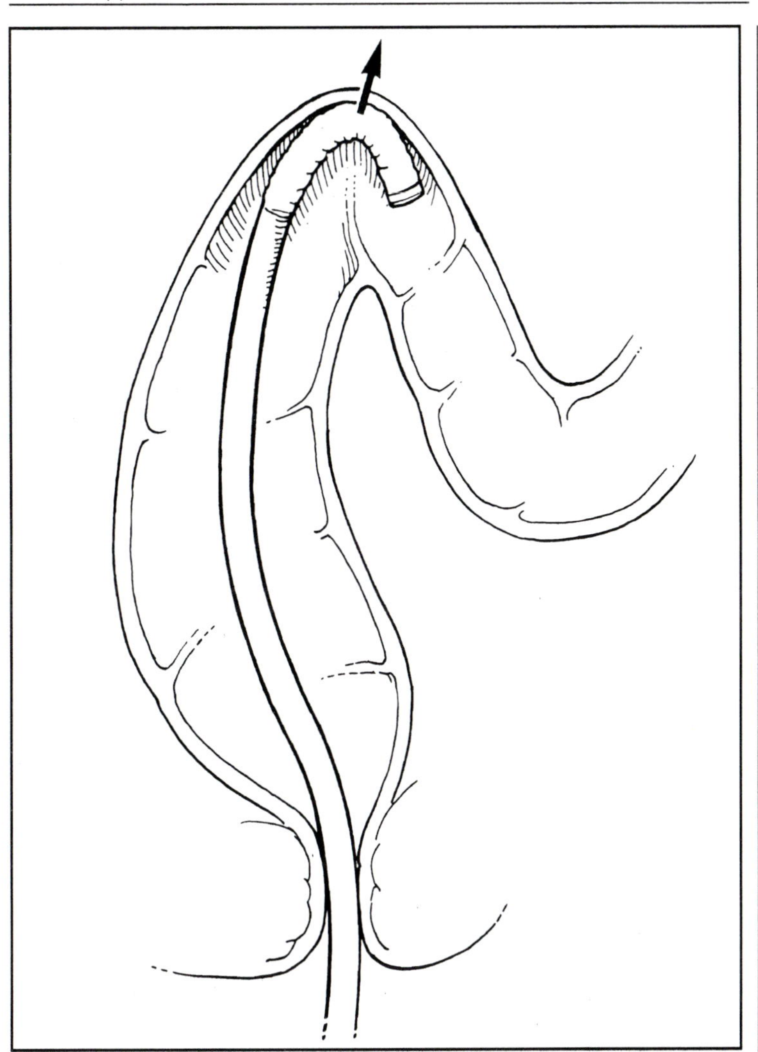

Figure 18.6. Application of pressure in areas of fixed resistance may lead to perforation (arrow) even when the lumen is in full view.

- Perforations may present immediately or late. Patients should generally be managed surgically, except in some cases where the perforation is recognized late and the evidence of peritoneal irritation appears localized.

- Post polypectomy syndrome
  - A transmural burn of the colon related to therapeutic cautery, without free perforation, can result in the **post polypectomy syndrome**.

- Patients present with pain, localized abdominal tenderness sometimes accompanied by peritoneal findings, and variable fever and white blood cell count.
- Perforation should be ruled out. Patients are managed by being made NPO, hydration, and intravenous antibiotics directed towards anaerobes and gram negative aerobes.
- The syndrome typically resolves over 1-3 days.

- Post polypectomy bleeding
  - **Post polypectomy hemorrhage** occurs in 0.3%-2.25% of patients undergoing polypectomy. Large polyp sizes a risk factor for post- polypectomy hemorrhage.
  - Hemorrhage may occur immediately after polypectomy or may be delayed for up to three weeks.
  - Immediate arterial hemorrhage can be managed endoscopically by cautery of the bleeding point using multipolar electrocautery or a heater probe, which may be proceeded by injection of epinephrine diluted 1:10,000 in normal saline using a sclerotherapy catheter. In the case of pedunculated polyps, immediate hemorrhage may be controlled by regrasping the polyp stalk firmly with the polypectomy snare and holding it for 10-15 min.
  - Most delayed bleeding stops with conservative therapy and supportive care. Delayed hemorrhage with continued passage of bloody stools after polypectomy should be managed by repeat colonoscopy with injection of dilute epinephrine (1:10,000 in normal saline) into the bleeding site using a sclerotherapy catheter and/or treatment of the bleeding point with the multipolar cautery probe or the heater probe.
- Cardiopulmonary complications. Either sedation or discomfort related to instrument passage may precipitate myocardial ischemia or cardiopulmonary collapse. Oxygen, and equipment and medications to provide cardiopulmonary resuscitation and hemodynamic support must be available during colonoscopy.
- Vasovagal reactions. Sedation or discomfort or both may result in **vasovagal reactions** presenting as bradycardia, hypotension, diaphoresis, or a combination of these. Atropine and general supportive care are of value in correcting the reaction.
- Oversedation
  - Combinations of benzodiazepines and narcotic analgesics are in widespread use for the production of sedation and analgesia during colonoscopy.
  - These medications may result in oversedation, presenting as inadequate patient responsiveness, hypotension, respiratory depression or hypoxemia.
  - Treatment consists of Trendelenburg position, fluid administration for hypotension, and supplemental oxygen for hypoxemia. The use of the reversal agents naloxone (Narcan 0.4 mg, 1 or 2 ampules intravenously) or flumazenil (Romazicon, 2 cc intravenously over 1 minute, repeated every several minutes as needed) may be necessary.
  - Administration of Narcan to patients on narcotics chronically can precipitate acute narcotic withdrawal, including pulmonary edema. Administration of flumazenil is contraindicated in patients with seizure disorders.

Technique

- Preprocedure assessment
- A history and examination must be recorded prior to colonoscopy. The history should record the indication for the procedure, the patient's other past and active medical problems, previous abdominal surgeries, and current medications and allergies.
- Particular note of the cardiopulmonary history and history of renal insufficiency, seizure disorder, and previous reactions to sedatives and analgesics.
- The examination must include the vital signs, a description of mental status, cardiopulmonary and abdominal findings.
- Monitoring
  - All patients should have continuous monitoring of blood pressure, pulse, and oxygen saturation.
  - One individual should continuously monitor the patient's level of consciousness and comfort level.
- Sedation
  - Sedation is usually used for colonoscopy in the United States.
  - Male gender, absence of abdominal pain, and increasing age are predictors of patients willing to consider colonoscopy without sedation.
  - Patients motivated to undergo colonoscopy without sedation can be successfully colonoscoped by experts without sedation in > 90% of cases.
  - The agents most commonly used for sedation for colonoscopy in the United States are benzodiazepines and narcotic analgesics.
  - Light conscious sedation is generally adequate to perform colonoscopy.
  - Midazolam (Versed) is more than three times more potent than diazepam (Valium). These drugs produce excellent sedation and desirable amnesia.
- Insertion
  - Digital rectal examination is performed prior to every colonoscopy.
  - The instrument should be guided whenever possible under direct visualization to the cecum. Direction of the lumen is determined by the circular shape of colonic folds, shadowing, and the course of the taenia.
  - Pain is avoided during colonoscopy by keeping the instrument shaft straight. Formation of loops should be avoided, and when necessary loops should be created only transiently and promptly reduced. Good technique utilizes regularly pulling back on the instrument to reduce looping, generally with right hand rotation (torque) on the instrument shaft.
  - The patient is typically positioned left lateral decubitus for insertion. Instrument advancement may be facilitated by changing to the supine position, particularly for crossing the transverse colon.
  - Right lateral decubitus position useful for passing the hepatic flexure or intubating the cecum from the mid-right colon or ileocecal valve.
  - Manual abdominal pressure placed by the assistant can help to resist loop formation and facilitate forward movement of the instrument.
  - The instrument should never be passed against fixed resistance, in which forward pushing with the instrument shaft is not accompanied by a sensation of the colon "giving way" in response to pressure. This rule holds even when the lumen is in full view (Fig. 18.6).

- Passage of the standard colonoscope may be difficult or impossible in patients with fixation of the sigmoid colon related to adhesions or previous diverticulitis, in patients with strictures or narrowing. In these cases, instrument passage will often be possible with use of a pediatric colonoscope or an upper endoscope.
- In rare instances of marked looping of the instrument, passage can be accomplished by use of an external straightener tube passed over the straightened instrument and into the sigmoid colon. Internal stiffening devices passed down the instrument channel are also available.
- Withdrawal and inspection involves a slow meticulous process of examination behind haustral folds, accompanied by adequate distention (Fig. 18.7).

• **Polypectomy**
- Endoscopic removal of all pedunculated polyps regardless of size is generally possible for experts. Sessile polyps are appropriate for removal if they occupy less than 30% of circumference of the bowel and do not cross over two adjacent haustral folds.
- **Cold forceps** may be used to remove polyps 1-3 mm in size with a single bite or in piecemeal fashion.
- **Cold snaring**, which involves placing a loop around the base of the polyp and guillotining the polyp base without cautery, can be used for polyps < 5 mm in size.
- **Hot forceps** can be applied to polyps < 7 mm in size. Anecdotal reports suggest that the use of hot forceps is accompanied by an unexpectedly high incidence of perforation and delayed bleeding. However, this remains unproven. Hot forceps are less safe in the right colon than in the left colon. They are associated with a 16% incidence of residual adenoma on subsequent examination.
- Hot forceps are used by placing the forceps on the tip of a polyp, tenting the polyp into the lumen, and applying monopolar cautery which is allowed to spread down the polyp edges to destroy the polyp. The biopsy specimen includes only the tip of the polyp which is protected from cautery by being in the forceps. The proper cautery effect has been described as producing a "Mount Fuji" appearance as the cautery burn is suggestive of snow spreading down the sides of a mountain from the top (Fig. 18.8 a, b). Hot forceps are generally considered the easiest cautery instruments to apply to small polyps and the hardest with which to control the cautery burn.
- **Snare cautery** is usually accomplished using monopolar wire loops. They are appropriate for any sessile or pedunculated polyps but tiny sessile polyps may be impossible to grasp with a snare. Experts vary in their use of pure coagulation versus blended (mixture of coagulation and cutting current) current. Most experts use pure low power coagulation current. Cautery is applied prior to manual transection of the polyp. The cautery results in desiccation of the tissue and easy manual transection.
- **Submucosal saline injection** is used to facilitate the removal of flat neoplastic lesions from the colon (Fig. 18.9). It may also reduce the risk of perforation associated with endoscopic resection of large sessile lesions, but this remains unproven. A sclerotherapy catheter is used to inject large amounts of submucosal saline (3 to 50 cc) into the submucosal space under the polyp. The polyp is then resected using a monopolar electrocautery snare.

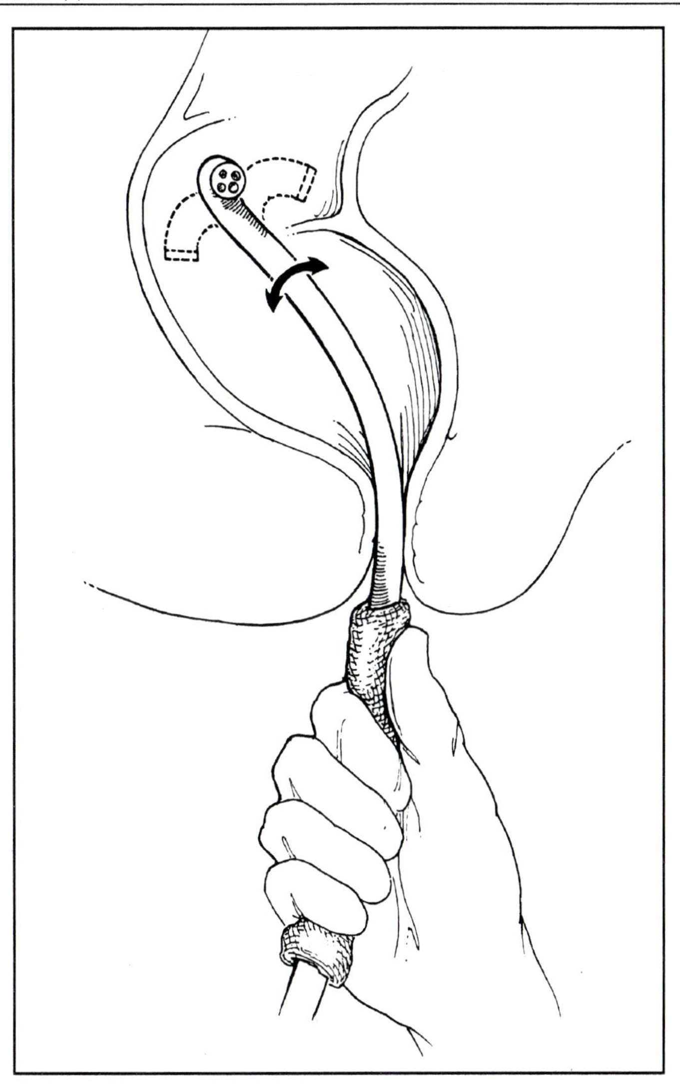

Figure 18.7. Careful withdrawal requires viewing the proximal sides of haustral folds, rectal valves, flexures and the ileocecal valve.

- Piecemeal polypectomy refers to the removal of a sessile lesion in two or more pieces, rather than a single grasping by the polypectomy snare. Piecemeal removal increases the safety of removal of large polyps and is a practical necessity for most large sessile lesions.
- **Dye spraying** (also called chromoendoscopy) involves spraying of materials such methylene blue, indocyanine green, or cresyl violet on the surface of

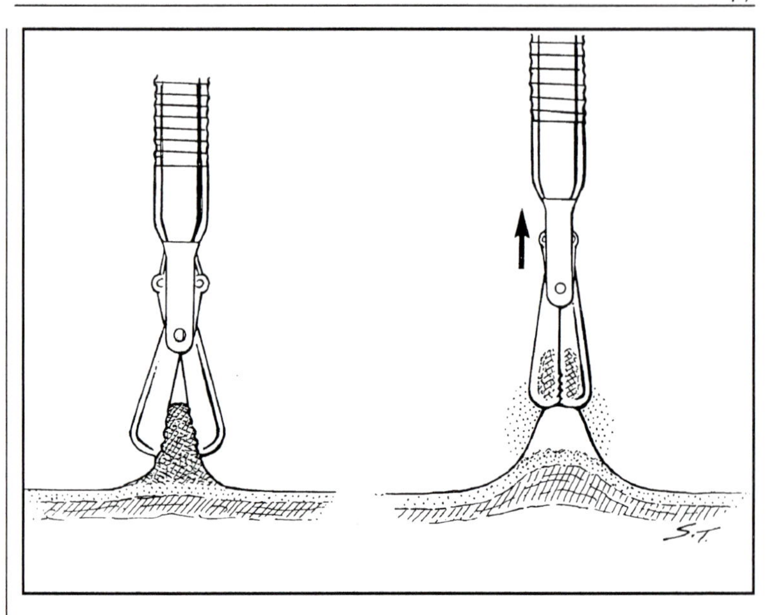

Figure 18.8. Correct application of hot forceps. The tip of the polyp is grasped with the forceps A. The polyp is pulled up or "tented" into the lumen and cautery is applied. The cautery spreads out from the forceps tip to destroy the polyp, while the biopsy sample is preserved B. Proper application requires that the cautery burn not spread excessively onto normal tissue.

the mucosa. In the area of polypectomy, it usually is used to facilitate identification of the subtle flat edges of large sessile polyps.

- **Tattooing** involves submucosal injection of a color marker to identify the lesion for subsequent surgical resection or reevaluation by repeat endoscopy. For practical purposes, the only useful colonic tattoo is sterile carbon particles (India ink).

• Destruction of vascular malformations
  - **Vascular malformations** in the colon are typically located in the cecum or ascending colon.
  - Endoscopic destruction is best accomplished with instruments that produce relatively shallow depth of penetration because of the thinness of the right colon wall. Thus monopolar cautery and Nd:YAG laser are less appropriate tools for destruction of vascular malformations in the right colon.
  - The use of multipolar coagulation, KTP laser, argon plasma coagulation, and the heater probe, are all appropriate tools for destruction of arteriovenous malformations.
  - Radiation proctitis generally affects the rectum or rectosigmoid colon. Bleeding from radiation proctitis can be controlled endoscopically in nearly all cases and endoscopic therapy is the treatment of choice. Options include Nd:YAG laser, argon plasma coagulation, multipolar electrocautery, and application of topical formalin.

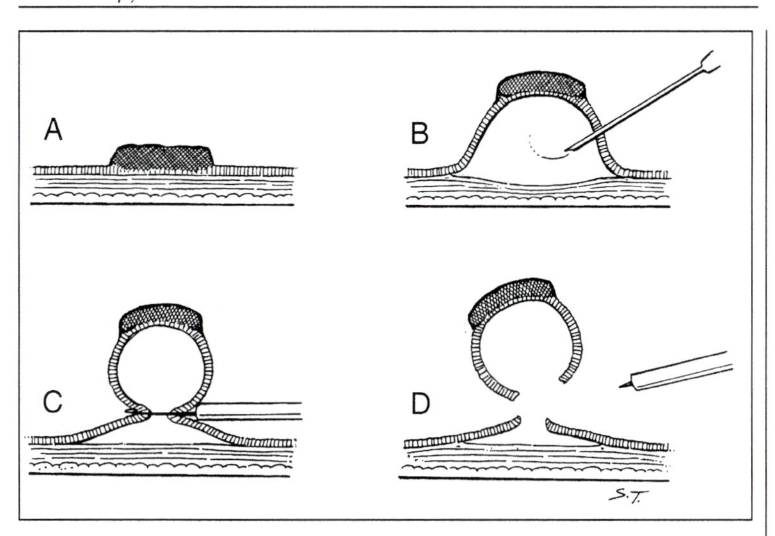

Figure 18.9. Submucosal saline injection on polypectomy. A. Sessile polyp. B. Injection of saline into the submucosal space. C. Application of the snare. D. Immediately post snare resection.

- Dilation of colonic strictures
  - Benign strictures may occur at the site of ileocolonic anastomosis, colocolonic anastomosis, and inflammatory bowel disease. Distal colonic strictures can be dilated using polyvinyl chloride flexible bougies passed over an endoscopically positioned guidewire (Savary system).
  - Through the scope (TTS) dilators or pneumatic balloons passed over an endoscopically placed guidewire were also effective.
  - Sequential dilation to diameters of 20-25 mm can generally be accomplished safely with a low risk of perforation. Injection of triamcinolone (up to 80 mg) into the stricture, after performance of dilation, may help to prevent recurrence in strictures that have a demonstrated the tendency to recur. However, this remains unproven.

Findings

- **Colorectal cancer** is the second leading cause of cancer death in the United States, with more than 130,000 new cases and 55,000 deaths per year.
  - The predominant risk factor for colorectal cancer is increasing age, and the lifetime incidence in men and women is nearly equal.
  - Colorectal cancer may produce symptoms of low gastrointestinal bleeding, either nonemergent or acute. Altered bowel habit and abdominal pain usually occur only in advanced disease. Colorectal cancers develop and grow in a asymptomatic phase for months to years before producing symptoms.
  - Approximately 40% of colorectal cancers arise proximal to the splenic flexure and approximately 50% are not detected by routine flexible sigmoidoscopy. 75% of patients with cancer proximal to the splenic flexure have no adenomatous polyps or cancer distal to the splenic flexure. Thus, 30% of all

patients with colorectal cancer will have a negative screening flexible sigmoidoscopy.

- The overall sensitivity of colonoscopy for detection of colorectal cancer is 95%. Nongastroenterologists are five time more likely than gastroenterologists to miss a colorectal cancer during colonoscopy.
- Maximization of the sensitivity of colonoscopy requires a standard of 90% or higher cecal intubation rates by colonoscopists. Careful inspection of blind areas proximal to the ileocecal valve, flexures, and rectal valves is essential.
- Cancer is identified at colonoscopy as an irregular mass lesion which may be sessile, polypoid, annular, or depressed and ulcerated. Ulcerations, irregularity, and firm texture favor cancer over benign tissue.
- Multiple biopsies should be taken of all suspicious lesions. Cytologic brushings are also recommended by some clinicians.
- Endoscopic resection of an apparently benign polyp may be followed by pathologic interpretation of invasive cancer. Endoscopic resection is considered sufficient for cure in pedunculated malignant polyps (the risk of metastasis from colorectal cancer is less than the risk of death from surgical resection) if the cancer is well or moderately differentiated, not invading vascular channels, and not approaching histologic resection line (cautery burn) to a distance closer than 1 or 2 mm.
- Some experts recommend surgical resection following endoscopic removal of a sessile malignant polyp. However, other experts believe endoscopic resection is adequate if the above histologic criteria for pedunculated polyps are met and the endoscopist is confident of a complete resection.

• **Adenomas** are present in 25-40% of average-risk adults age 50 or older in the United States. They are more common in age matched males than females and somewhat more common in persons with a family history of either cancer or adenoma.
  - More than 90% of adenomas are < 1 cm in size. The risk of cancer in an adenomatous polyp increases with the size of the adenoma and the extent of villous histology.
  - All adenomas are dysplastic. Dysplasia is categorized pathologically as low-grade or high-grade. The term "carcinoma in-situ" is inappropriate (high-grade dysplasia is preferred), since the risk of metastasis in patients with high-grade dysplasia or cancer in the lamina propria morphologically (intramucosal adenocarcinoma) is zero, and such lesions should be treated clinically as benign polyps.
  - Adenomas are classified histologically as villous if they have greater than 75% villous elements (frond-like glandular pattern) and tubular if they have more than 75% tubular elements (organized glandular structure). Mixed polyps are called tubulovillous. Patients with only small tubular adenomas may not have a lifetime risk of colorectal cancer greater than the general population, whereas patients with villous or tubulovillous histology clearly do have increased risk.
  - Adenomatous polyps cannot be differentiated from nonneoplastic polyps such as hyperplastic polyps by their endoscopic appearance alone. However, large size and red color favor adenomatous histology. Current practice in the United States is to remove all polyps identified during colonoscopy and submit them for pathologic evaluation.

- Patients with adenomatous polyps are at increased risk of subsequent colorectal cancer and should generally undergo surveillance (see indications).

- **Diverticulosis**
  - Asymptomatic diverticulosis affects half or more of Americans with increasing age.
  - Diverticulosis is the most common cause for massive hemorrhage in lower gastrointestinal tract. Small number of patients with ongoing diverticular hemorrhage have been successfully managed by colonoscopy, identification of the bleeding diverticulum, followed by injection of epinephrine and/or electrocautery of the bleeding point.
  - The presence of edematous folds, sometimes accompanied by visualization of a diverticulum with ulceration, inflammation, and mucopus exuding from the diverticulum favors the diagnosis of acute diverticulitis. Full colonoscopy should proceed cautiously in the setting of acute diverticulitis in order to avoid free perforation of the infected diverticulum related to air insufflation.
- **Ulcerative colitis**
  - Colonoscopy and biopsy is the most sensitive way to determine the extent and activity of ulcerative colitis.
  - Patients with proctitis or rectosigmoiditis may demonstrate inflammation in the area of the cecum, referred to as the "cecal patch". The clinical significance of a cecal patch remains undetermined.
  - The severity of ulcerative colitis is best judged by clinical criteria and typically categorized as mild, moderate or severe. There is not a perfect correlation between endoscopic findings and clinical status. Thus, some patients with mild endoscopic findings have severe clinical symptoms, and approximately 10% of patients with severe endoscopic ulcerative colitis are asymptomatic. Nevertheless, there is a general correlation between endoscopic findings and symptoms, as well as endoscopic healing and clinical remission.
  - Findings of mucosal atrophy and distortion of the vascular pattern are seen in quiescent (inactive) ulcerative colitis.
  - Mild disease activity is characterized by erythema, surface granularity (related to edema), distortion of the vascular pattern, and friability (the mucosa bleeds easily when touched).
  - Moderate activity consists of worsening of the findings of mild activity plus the presence of mucopus in the lumen, and ulcerations < 5 mm in size.
  - Severe endoscopic ulcerative colitis is characterized by the above findings plus ulcers > 5 mm in size and evidence of spontaneous hemorrhage (blood in the lumen).
  - Patients with long standing ulcerative colitis incur an increased risk of colorectal cancer and should be offered interval surveillance colonoscopy (see indications).
  - The effectiveness of **surveillance colonoscopy** in the prevention of colorectal cancer death in chronic ulcerative colitis has not been studied in a randomized controlled trial, and several lines of evidence suggest that it is less than perfectly effective. Patients should be instructed regarding the limitations of surveillance colonoscopy in ulcerative colitis and offered the alternative of prophylactic proctocolectomy, particularly after 20-30 years of symptoms in panulcerative colitis.

- In order to obtain a 90% confidence that dysplasia has been detected if present, at least 33 biopsies must be taken during surveillance colonoscopy. A standard protocol involves four quadrant biopsies from the cecum followed by four quadrant biopsies every 10 cm. Biopsies should be directed toward any areas of velvety or villiform mucosa. Irregular mass lesions should be biopsied separately. Typical-appearing adenomatous polyps should be removed by snare polypectomy. Obvious pseudopolyps need not be biopsied. Biopsy technique should utilize jumbo forceps and "turn and suction" technique. In the latter technique, the open forceps are placed against the mucosa and suction is used to make the mucosa fall into the forceps before closing.
- If low or high-grade dysplasia is encountered, it should be confirmed by an expert in ulcerative colitis pathology, and when confirmed, the patient should be advised to undergo total proctocolectomy.
- Identification of a **dysplasia associated lesion or mass** (DALM) is an indication for colectomy. These lesions may be difficult to distinguish from routine adenomatous polyps, which are also mass lesions that are dysplastic. Sessile lesions with surface irregularity in shape or color or a positive p53 stain favors DALM rather than adenoma.
- Adenomas proximal to an involved segment or a reading of "indefinite for dysplasia" in biopsies from flat mucosa can be followed by continued endoscopic surveillance.
- The approach to patients with adenomatous polyps arising in areas of active or previously active ulcerative colitis is controversial. Some experts believe that if patients are age 50 or older with one or two adenomas in an involved segment they can be followed by continued surveillance. Others believe that all patients with adenomas in an involved segment of ulcerative colitis should undergo colectomy.

- Crohn's disease
  - The endoscopic features of **Crohn's disease** include skip lesions, noncircumferential involvement, deep, serpiginous longitudinal ulcers, cobble stoning, and fistula openings. In about 15% of patients with inflammatory bowel disease, it is not possible to distinguish ulcerative colitis from Crohn's disease.
  - Colonoscopy is useful in identifying the extent and severity of Crohn's disease and allows visualization of active disease in the terminal ileum.
  - Colonoscopy is very useful for evaluation of recurrent disease following resection of the terminal ileum and cecum. In such patients recurrences typically occur in the neoterminal ileum immediately proximal to an involved anastomosis.
  - Patients with Crohn's disease are generally managed by the clinical response to treatment. Endoscopic follow-up to document the response to treatment is not clinically useful.
  - Patients with long standing Crohn's colitis that have not been operated are at increased risk for development colorectal cancer. They should undergo colonoscopy with biopsies for dysplasia of all involved areas.

- Ischemic colitis usually presents in elderly patients who often have widespread atherosclerotic disease and may have suffered recent cardiovascular compromise. Presenting findings generally include abdominal pain, usually in the lower quadrants and often left-sided, diarrhea, overt or occult rectal bleeding, abdominal tenderness on examination, and thumb printing on plain abdominal x-rays (thickened edematous haustral folds in the involved segment).
  - Mesenteric angiography is not indicated in the diagnosis of ischemic colitis (It is critical for the diagnosis of mesenteric ischemia involving the small intestine).
  - Colonoscopy is the procedure of choice for diagnosis of ischemic colitis. Generally, the colonoscope should be passed at least proximal to the splenic flexure, as ischemic colitis typically occurs anywhere in the left colon from rectosigmoid junction to splenic flexure. Occasionally, ischemic colitis is located in the right colon. Right colonic ischemia should raise concerns regarding coincident mesenteric ischemia. Ischemic proctitis is rare and has been reported primarily in patients with extensive pelvic surgery resulting in devascularization of the rectum.
  - Colonoscopy is performed cautiously in suspected ischemic colitis. Findings which favor the diagnosis are dusty or purple color of the mucosa, submucosal hemorrhage with mucosa sloughing and bleeding, and nonspecific scattered ulcers of variable size, particularly if they are long and situated along one wall of the colon.
  - Treatment of the ischemic colitis involves bowel rest, antibiotics if there is fever or elevated white blood count, and careful serial abdominal examinations. Development of peritoneal findings indicates transmural infarction and is an indication for surgical resection of the involved colonic segment.
  - Ischemic colitis is most often followed by complete healing but may present with recurrent acute episodes or may be followed by stricture formation.
- Radiation proctitis. Symptomatic **radiation proctitis** may present months to years after the radiation course.
  - Bleeding from radiation injury to the colon is invariably accompanied by endoscopic identification of telangiectasia like lesions, usually in the rectum or rectosigmoid colon.
  - Medical therapies such as steroid enemas, 5-aminsalicylate enemas, etc. are of no proven value and generally are completely ineffective in control of bleeding from radiation injury.
  - Endoscopic therapy is the treatment of choice for bleeding radiation injury (see destruction of vascular lesions – above).
- Solitary rectal ulcer syndrome (SRUS) typically presents with hematochezia and symptoms of disorder defecation including tenesmus, diarrhea, and outlet obstruction constipation.
  - Endoscopic findings can include polyps, patchy erythema, or ulceration.
  - The diagnosis is based on distinct histologic features obtained from multiple biopsies or endoscopic resection of a polyp.
  - The etiology of SRUS in some patients is believed to be from mucosal prolapse. However, the etiology is established and patients are often refractory to treatments. Referral to a gastroenterologist or colorectal surgeon is indicated for SRUS.

Outcomes

- Colorectal cancer in patients with adenomas. In two cohorts of patients with adenomas who underwent clearing colonoscopy, removal of all colonic polyps was associated with a dramatic reduction in the incidence of colorectal cancer and near complete protection from colorectal cancer and mortality. Colonoscopy is currently the most powerful visceral cancer prevention technique in clinical medicine.
- Colorectal cancer in ulcerative colitis. In ulcerative colitis uncontrolled data suggest that surveillance colonoscopy has a substantial protective effect from colorectal cancer mortality in chronic ulcerative colitis patients. However, this remains unproven, and there are also multiple cases of colorectal cancer deaths occurring in patients adhering to surveillance programs.

Selected References

1. Rex DK. Colonoscopy: A review of its yield for cancers and adenomas by indications. Am J Gastroenterol 1995; 3:253-265.
2. Rex DK, Cutler CS, Lemmel GT et al. Colonoscopic miss rates of adenomas determined by back-to-back colonoscopies. Gastroenterology 1997; 112:24-28.
3. Rex DK. Rahmani EY, Haseman JH et al. Relative sensitivity of colonoscopy and barium enema for detection of colorectal cancer in clinical practice. Gastroenterology 1997; 112:17-23.
4. Waye JD. New methods of polypectomy. In: Sivak MV, Rex DK. Gastrointestinal Endoscopy Clinics of North America. Philadelphia: W. B. Saunders Company, 1997:413-422.
5. Winawer SJ, Zauber AG, Ho MN et al. Prevention of colorectal cancer by colonoscopic polypectomy. N Engl J Med 1993; 329:1977-1981.
6. Winawer SJ, Zauber AG, O'Brien MJ et al. Randomized comparison of surveillance intervals after colonoscopic removal of newly diagnosed adenomatous polyps. N Engl J Med 1993; 328:901-906.
7. Rex DK. Instruments and equipment. In: Raskin JB, Nord HJ. Colonoscopy Principles and Techniques. New York: IGAKU-SHOIN Medical Publishers, Inc., 1995:17-33.
8. Rex DK. Acute colonic pseudo-obstruction (Ogilvie's Syndrome). Gastroenterologists 1994; 2:233-238.

CHAPTER 19

ERCP—Introduction, Equipment, Normal Anatomy

Gerard Isenberg

Introduction

Endoscopic retrograde cholangiopancreatography (ERCP), first reported in 1968,[1] encompasses various procedures in the diagnosis and treatment of diseases of the biliary tree and pancreas. Using contrast dye injected via a small catheter, the common bile duct, the intrahepatic ducts, the cystic duct and gallbladder as well as the pancreatic ductal system can be visualized under fluoroscopy. Although it seems straightforward, ERCP can be technically challenging because of anatomic variants, postoperatively altered anatomy, and pathologic changes. Depending on the disease, various diagnostic (including brush cytology and biopsy) and therapeutic (including endoscopic sphincterotomy, basket extraction of stones, and stent placement) measures can be performed. ERCP should only be performed by those capable of proceeding with therapeutic interventions.

Indications[2] and Contraindications

- **Diagnostic ERCP** is generally indicated in:
  - Evaluation of the jaundiced patient suspected of having biliary obstruction.
  - Evaluation of the patient without jaundice whose clinical presentation and biochemical or imaging data suggests biliary tract or pancreatic disease.
  - Evaluation of signs or symptoms suggesting pancreatic malignancy when results of indirect imaging [i.e., ultrasound (US), computerized tomography (CT), or magnetic resonance imaging (MRI)] are equivocal or normal
  - Evaluation of recurrent or moderate to severe pancreatitis of unknown etiology.
  - Preoperative evaluation of the patient with chronic pancreatitis and/or pseudocyst.
  - Evaluation of the sphincter of Oddi by manometry.
- Diagnostic ERCP is generally not indicated in:
  - Evaluation of abdominal pain of obscure origin in the absence of objective findings which suggest biliary tract or pancreatic disease.
  - Evaluation of suspected gallbladder disease without evidence of bile duct disease.
  - As further evaluation of pancreatic malignancy which has been demonstrated by US or CT unless management will be altered.
- Therapeutic ERCP is generally indicated for:
  - Endoscopic sphincterotomy (ES) in choledocholithiasis and in papillary stenosis or sphincter of Oddi dysfunction causing significant disability

Gastrointestinal Endoscopy, edited by Jacques Van Dam and Richard C. K. Wong.

- To facilitate placement of a biliary stent or balloon dilatation of a biliary stricture
- Sump syndrome
- Choledochocele involving the major papilla
- Ampullary carcinoma in patients who are not candidates for surgery.
- Stent placement across benign or malignant strictures, biliary fistula, postoperative bile leak, or in "high risk" patients with large, unremovable common duct stones.
- Balloon dilatation of biliary stricture.
- Nasobiliary drain placement for prevention or treatment of acute cholangitis or infusion of chemical agents for common duct stone dissolution, for decompression of an obstructed common bile duct, or postoperative biliary leak if stent placement is unsuccessful or unavailable.

Contraindications

- Absolute
  - Recent acute pancreatitis unrelated to gallstones.
  - Medically unstable patient.
  - Uncooperative patient.
- Relative
  - Bleeding diathesis (correct coagulopathy and/or thrombocytopenia).
  - Pregnancy
  - The risks and benefits of performing fluoroscopy and endoscopy with the potential side effects of radiation exposure and medication use needs to be carefully balanced with the clinical need for ERCP and possible therapy.
  - Previous contrast reaction.
  - The use of low ionic or nonionic contrast agents should be employed. Depending on institutional preference, a protocol utilizing diphenhydramine, an H_2 blocker (i.e., ranitidine), and prednisone may be used for pretreatment in the case of previous contrast reaction.
  - Residual barium from previous examination which will obscure contrast injection. A scout film of the abdomen performed prior to sedation of the patient can confirm the absence of barium. It will also identify artifacts (i.e., radiopaque items on clothing and surgical clips) and calcifications (i.e., those in the pancreas, lymph nodes, or on ribs) which may be superimposed upon the fluoroscopic field of interest and can lead to misinterpretation of the ERCP films.
  - Recent myocardial infarction or significant arrhythmia.
  - Again, the risks and benefits of proceeding with ERCP and potential therapy need to be balanced with the clinical situation.

Equipment, Endoscopes, Devices and Accessories

- A side-viewing duodenoscope enables excellent visualization of the stomach and proximal duodenum to the papilla of Vater. The video endoscope with an electronic "chip" allows a brilliant television image to be displayed. Generally, a 5 F Teflon catheter with graduated tip markings is used to cannulate the papilla. High-resolution fluoroscopic equipment with image intensification is needed to provide high-quality imaging and radiographs. The radiology table should tilt to permit oblique and erect films. To minimize radiation exposure the endoscopy team will wear lead wraparound aprons and thyroid collars. Current occupational guidelines permit exposure of personnel to 5 rem (roentgen equivalent man) per year. Electrocautery units and a complete range of endoscopic accessories will be needed should therapeutic endoscopy be necessary.

Endoscopes

- Diagnostic side-viewing duodenoscope. Generally, this instrument (1012.5 mm in outer diameter with a 3.2-3.8 mm working channel) is used in cases in which therapeutic maneuvers will not be necessary (i.e., evaluation of recurrent pancreatitis).
- Therapeutic side-viewing duodenoscope. This instrument measures 12.5-14.5 mm in diameter and with its larger 4.2 mm working channel is able to accommodate a variety of accessories needed for therapeutic maneuvers.
- Forward-viewing endoscope. Occasionally, this instrument will be used in post-surgical changes of the stomach and duodenum (i.e., Billroth II) to allow for cannulation of the papilla of Vater.

Devices

- **Electrocautery unit**. A variety of electrosurgical monopolar and bipolar generators is commercially available. Most endoscopy units use the same generators for polypectomy which offer pure cutting, pure coagulating, and blended modes, for endoscopic sphincterotomy (ES). The optimal current for ES has not been determined, but most centers prefer the blended current. Using snare polypectomy techniques, the device can also be used in the treatment of ampullary tumors.
- **Mechanical lithotripter**. This device is used to crush gallstones within the biliary tree by mechanical shortening of a basket catheter surrounding the stones.
- **Laser unit**. Few endoscopy units have the capability to perform laser lithotripsy of stones and to use photodynamic therapy in the treatment and palliation of tumors of the biliary system.
- **Direct cholangioscopy/pancreatoscopy**. Mother-daughter endoscopes allow direct visualization into the bile and pancreatic ducts. The “daughter” scope is a small caliber endoscope that can be inserted into the working channel of a separate therapeutic “mother” duodenoscope and has a separate processor and imaging system. Small visually-directed biopsies may be obtained in this fashion.
- Catheter-based **endoscopic ultrasound probes**. Excitement has been generated regarding the capability of these probes to stage ampullary carcinomas and cholangiocarcinomas and to distinguish air bubbles from gallstones. These probes, some of which are wire-guided, are available at only a few centers.

Accessories

- Catheters. An assortment of cannulating **catheters** is available. They differ primarily in the shape of the tip. The tip may be metal or nonmetal. Catheters are used to opacify the biliary and pancreatic ducts with radiopaque contrast to allow for fluoroscopic viewing. Most catheters have three, 3 mm etched markings at the distal tip. The standard or slightly tapered catheters accept a 0.035 inch guidewire (described below). If choledocholithiasis is suspected, it is generally wise to proceed with initial cannulation with a papillotome (described below). When biliary cannulation is unattainable with a standard catheter, a fine-tapered catheter may be tried. The main disadvantages of this type of catheters are that the sharp tip can more easily cause trauma to the papilla, submucosal injections occur more frequently, the resistance to injection is increased, and the catheter can only accommodate smaller guidewires (i.e., 0.018 and 0.025 in). Standard-tapered, ultra-tapered, and needle-tipped catheters may be used for minor papilla cannulation (i.e., as in suspected pancreas divisum).

- **Papillotomes** (sphincterotomes). An enormous variety of papillotomes is commercially available. The most commonly used is the original Erlangen pull-type bowstring design.

    The main differences between papillotomes are the length of the exposed cutting wire (typically 20-30 mm), the number of additional lumens (either one or two for separate guidewire and contrast injection), and the length of the nose extending beyond the cutting wire (typically 58 mm with a range of up to 50 mm). Prior partial gastrectomy with a Billroth II anastomosis changes the orientation of the papilla such that the bile duct enters at the 6 o'clock position. Specially designed papillotomes have been developed for such cases.
- Guidewires. A vast array of **guidewires** is also available with a variety of sizes, including 0.035", 0.025", and 0.018" in diameter. Most guidewires have a hydrophilic coating. Many guidewires have been developed with a special hydrophilic coating that allows the guidewire to remain in place to assure access while an ES is carried out. These are called "protected" guidewires and prevent transmission or dissipation of the electrical current.
- Balloons. Extraction **balloons** are used in the endoscopic removal of biliary or pancreatic stones and are commercially available in a number of sizes and balloon volumes. Balloon dilators are designed to expand the intraductal lumen in areas of strictures. These also come in a variety of sizes and balloon diameters (inflated).
- **Baskets**. Constructed of braided wire, these devices are useful in the extraction of biliary stones.
- Stents. There is an extraordinary number of **stents** available in varying lengths, diameters, side hole and flap designs, configurations (straight versus curved and pigtail) and materials (plastic, metal, and Teflon). The selection of the stent depends on the clinical situation. Conventional plastic stents, which are inexpensive, typically develop occlusion by bacterial biofilm after 3 to 6 months and require replacement to maintain patency if still clinically indicated. Expandable metallic stents, which are more expensive, are typically used to palliate malignant biliary strictures as they have greater longevity.

Technique

- To convey the technical nuances of performing ERCP is outside the scope of this handbook, or for that matter, any textbook. Whereas the American Society for Gastrointestinal Endoscopy (ASGE) has established a minimum of 100 ERCPs (75 diagnostic and 25 therapeutic), it is unlikely that any practitioner who has performed less than 200 ERCPs during training is competent enough to attain greater than an 85% success rate in all situations where an ERCP is required.[3]

History

- A previous history of contrast reaction should be elicited as noted above.
- In women of childbearing age, it is imperative that pregnancy be addressed. If necessary, a urine pregnancy test should be performed.
- If the patient has a pacemaker or an automatic implantable cardioversion device (AICD), consideration should be made to consult cardiology in the event that the pacemaker or AICD needs to be turned off during electrocautery to prevent inadvertent programming problems or firing.
- A surgical history, particularly regarding operations of the stomach and small intestine (i.e., Billroth I or II), should be elicited as this may impact on the choice of endoscope used for ERCP.
- If a patient is on insulin, generally, one-half of the usual dose is given on the morning of the exam to prevent hypoglycemia.

Laboratory Data

- Most often, patients have had a chemistry panel, including glucose, blood urea nitrogen (BUN), and creatinine, a complete blood count (CBC), including a platelet count, and prothrombin and partial thromboplastin time (PT/PTT) measured as part of their workup leading to an ERCP.
- If a bleeding history is elicited, a platelet count and PT/PTT is necessary, particularly if brush cytology, biopsy, and therapeutic maneuvers are considered.
- Coagulation status
- Patient on heparin. Heparin should be stopped for 4 hours prior to the procedure to allow the activated partial thromboplastin time (aPTT) to normalize. Heparin may be restarted 6 hours after completion of the procedure. If an endoscopic sphincterotomy is performed, consideration of a longer delay in restarting heparin should be considered.
- Patient on warfarin. **Warfarin** is generally held prior to the procedure to allow partial normalization of the PT. Alternatively, fresh frozen plasma (FFP) should be given prior to the procedure. A PT of less than 15 seconds and an Internationalized Normal Ratio (INR) of less than 1.4 are desirable, especially if an ES is considered. Vitamin K should be avoided as this makes re-anticoagulation with warfarin difficult.

Patient Preparation

- Informed consent is obtained. The procedure, its benefits, its potential complications, and alternatives are discussed in detail.
- Aspirin and other nonsteroidal antiinflammatory (NSAIDs) medications should be withheld for several days before ERCP and also after ERCP, if ES is performed.
- Broad-spectrum intravenous antibiotics are administered if cholangitis, biliary obstruction, or pancreatic pseudocyst is suspected. They are also given in certain medical conditions, such as mitral valve prolapse with mitral regurgitation, prosthetic heart valve, a history of endocarditis, a systemic pulmonary shunt, or a synthetic vascular graft within the last year.
- Generally, the patient should have nothing by mouth (NPO) except for medications for 8 hours prior to ERCP.
- Insulin dosages should be adjusted as previously noted.
- An intravenous line should be placed to allow for sedatives and hydration.

- The patient is placed in left lateral position with his/her left arm behind his/her back on the fluoroscopy table. During the procedure, the patient is moved to the prone position with abdomen down for precise definition of ductal anatomy.
- Continuous monitoring of the patient's oxygen saturation, respiratory rate, heart rate, blood pressure, and responsiveness is employed throughout the procedure and postprocedure period until the patient returns to his/her baseline status.
- Local oropharyngeal anesthesia to suppress the gag reflex is obtained with any one of the varieties of topical sprays available (i.e., Cetacaine spray or Hurricane spray).
- Conscious sedation is usually achieved with a combination of intravenous medications (i.e., meperidine, midazolam, diazepam, and haloperidol), titrated to the desired effect. Less than 4% of patients may require general anesthesia, including those with mental retardation, a previous failed attempt with conscious sedation, and tolerance to medications secondary to substance abuse or narcotic use for pain.
- Oxygen as delivered by nasal cannula and intravenous hydration (i.e., normal saline) may be administered as clinically indicated.
- **Endoscopic intubation**. For the beginner, the two major challenges are passage of the scope through the pylorus and properly lining up the ampulla of Vater. The duodenoscope is inserted into the mouth, and using indirect visualization, the esophagus is intubated. If attempts to intubate the esophagus are unsuccessful, consideration should be made to use a forward-viewing endoscope to identify the problem.

  Upon entering the stomach, the lesser curvature is first visualized. Insufflated air is used to distend the stomach and enhance the view. Slight downward tip deflection will usually offer a tubular view of the stomach. Careful inspection of the fundus and upper body of the stomach is made with retroflexion of the instrument. The endoscope is then pulled back. The tip of the endoscope is angled down and advanced through the body of the stomach. As the instrument is side-viewing, the 6 o'clock position in view is the field ahead of the tip of the instrument. The pylorus is approached, and upon positioning it in the middle of the viewing field, the tip of the instrument is deflected upwards with subsequent passage into duodenal bulb. With rightward rotation and forward pressure, the endoscope will pass into the second portion of the duodenum. The duodenoscope is then moved down to the distal second portion of the duodenum, and then with right and upward tip deflection and clockwise torque, the duodenoscope is pulled back in simultaneous motion until the papilla of Vater is visualized. The resultant straightening of the endoscope moves the instrument forward. The beginner often pulls the instrument too quickly without enough torque and finds the instrument back in the stomach. With practice, he/she will be able to perform this maneuver so that only 60-70 cm of the endoscope is inside the patient. Often, fluoroscopy can be used for beginners to show them the position of the instrument during the various maneuvers used to visualize the papilla. Withdrawing the endoscope in proper position leads to precise tip control for cannulation. The papilla of Vater is typically found on the posteriomedial wall of the middle third of the descending duodenum. The papilla can vary widely in size, shape, and appearance. Occasionally, it is necessary to push the instrument in to the distal second portion of the duodenum in the "long" position to identify the papilla. If attempts fail to locate the papilla, a careful search of the posterior medial wall from the third portion of the duodenum to the bulb is carried out in a slow and deliberate manner, looking for a longitudinal fold or a stream of bile.

- **Cannulation of the papilla.** Once the papilla is identified and a good position for cannulation is obtained, it is useful to lock the controls so that the endoscope position is maintained. Glucagon or atropine is usually given at this time for intestinal ileus. A catheter is then introduced through the channel of the duodenoscope. Using fine adjustments of the biplane directional controls and manipulation of the elevator, the tip of the catheter is introduced along the axis of the desired duct. Often, using slight body movements by the endoscopist will have successful effects on the orientation of the endoscope tip and axis for cannulation. Successful positioning takes considerable practice. Incorrect positioning of the endoscope at the papilla is often the reason for prolongation of the procedure time and failure of ERCP. Selective cannulation of the bile and pancreatic ducts is another hurdle for beginners. The endoscopist should resist the temptation to cannulate the papilla immediately after it is seen. Close endoscopic evaluation of the papilla is warranted. The movements of the endoscope that are necessary to successfully line up the endoscope in a proper position for cannulation are not entirely predictable and are often performed in a trial and error method. The bile duct usually descends steeply along the posterior wall of the duodenum and joins the papillary orifice in the upper left portion, whereas the pancreatic duct opens fairly horizontally into the inferior right region of the orifice. Thus, for bile duct cannulation, the catheter is directed in the 11 o'clock axis. Once the papilla is entered, it is often useful to lift the catheter tip upwards with the elevator toward the roof of the papilla. For pancreatic duct cannulation, the catheter is directed slightly rightward at the orifice in a 1 o'clock axis. Successful cannulation depends on mastering the approach to the papilla together with the fine movements of passing the catheter.
- **Contrast injection**
  - The most frequently used contrast agent is a 50-60% water-soluble iodinated contrast. If a bile duct is known to be large or if choledocholithiasis is suspected, a more dilute contrast is employed (25-30%) since the density of such a thick column prevents smaller stones from being seen.
  - Iodinated contrast is preferred over the non-iodinated kind since it gives better resolution and is less viscous. In patients with previous contrast reactions, low ionic or nonionic contrast with a pretreatment regimen as described previously should be used.
  - The catheter should be flushed free of air bubbles. Using fluoroscopy, a small amount of contrast is injected. Contrast in the distal portion of the either duct can disappear within seconds after injection; thus, it is important to closely observe the area being injected. The reason for being cautious with the initial injection is that the main duct of the ventral pancreas may be very small, and the entire system can be filled with just 12 cc of contrast.
  - If selective cannulation of the desired duct is unsuccessful, it is necessary to withdraw the catheter and ensure that the proper axis is being obtained. Occasionally, the catheter may be withdrawn a few millimeters with contrast injection before the duct is subsequently visualized.
  - Injection should be gradual, steady, and with careful fluoroscopic monitoring, particularly when observing the pancreatic duct. When the pancreatic

Table 19.1. Grading system for the major complications of ERCP and ES

	Mild	Moderate	Severe
Bleeding	Clinical (i.e., not just endoscopic) evidence of bleeding, hemoglobin	Transfusion (4 units or less), no angiographic intervention or surgery	Transfusion 5 units or more, or intervention (angiographic or surgical)
Perforation	Possible, or only very slight leak of fluid or contrast, treatable by fluids	Any definite perforation treated medically for 4-10 days	Medical treatment for more than 10 days, or intervention and suction for 3 days or less (percutaneous or surgical)
Pancreatitis	Clinical pancreatitis, amylase at least 3 times normal at more than 24 hours after the procedure	Pancreatitis requiring hospitalization of 4-10 days	Hospitalization for more than 10 days, or hemorrhagic pancreatitis, phlegmon or-

duct is seen to the tail, injection should be stopped to prevent acinarization due to over-injection and leading to possible postERCP pancreatitis.
- Care must be taken to avoid missing abnormal findings in the portion of the bile duct or pancreatic duct over which the endoscope lies.
- Spot films should be obtained using cassettes to allow for review. It is occasionally useful at the end of the procedure to have the patient lie on their back to allow further imaging of the biliary system.

Outcome

- Proficient endoscopists should achieve successful cannulation of the biliary system and pancreatic duct in over 95% of cases. If therapeutic measures are required, it should be performed at the same setting. Sphincterotomy and stone extraction performed by experienced endoscopists are successful in 85-90% of cases.[4]

Complications

- In addition to complications related to endoscopy itself (adverse medication reaction, bleeding, infection, and perforation), there are several complications unique to ERCP and ES. For ERCP alone, the overall complication rate is approximately 4-6% with a mortality rate of less than 0.4% in experienced hands. With the addition of ES, the overall complication rate is 10% with a mortality rate of 1%.[5] Generally, a grading system of mild, moderate, or severe is used to categorize complications (see Table 19.1).[6]
  - **Pancreatitis** is the most common complication, occurring in approximately 5% of patients. Sphincter of Oddi manometry and young age represent two of the most important risk factors. Other risk factors are related to difficulty in cannulating the bile duct. Asymptomatic hyperamylasemia occurs in up to 75% of patients undergoing ERCP, and such patients should not be construed as having clinical pancreatitis.
  - **Bleeding**, which is most often evident at the time of ES, occurs with a frequency of 13%. The majority of episodes can be managed endoscopically with local injection of 1:10,00 epinephrine, multipolar electrocoagulation, or completion of the ES to allow full retraction of the partially severed vessel.
  - **Cholangitis** following ERCP and ES develops in 13% of patients, although the risk is higher when cholangitis is present prior to the procedure.
  - **Retroperitoneal perforation** usually occurs when the ES incision extends beyond the intramural segment of the bile duct into the retroperitoneal space and is documented by the presence of extravasated contrast or retroperitoneal air. The risk is less than 1%.
  - **Recurrent choledocholithiasis** and/or cholangitis and papillary stenosis may occur after ES in approximately 10% (range, 415%) of patients based on long-term follow-up studies.[79]

Selected References

1. McCune WS, Shorb PE, Moscowitz H. Endoscopic cannulation of the ampulla of Vater: A preliminary report. Ann Surg 1968; 167:752-756.
2. Appropriate use of gastrointestinal endoscopy. Manchester: American Society for Gastrointestinal Endoscopy, 1992.
3. Jowell PS, Baillie J, Branch MS et al. Quantitative assessment of procedural competence: A prospective study of ERCP training. Ann Intern Med 1996; 125:983-989.

4. Cotton PB. Endoscopic management of bile duct stones (apples and oranges). Gut 1984; 25:58797.
5. Freeman ML, Nelson DB, Sherman S et al. Complications of endoscopic biliary sphincterotomy. N Engl J Med 1996; 335:909918.
6. Cotton PB, Lehman G, Vennes J et al. Endoscopic sphincterotomy complications and their management: An attempt at consensus. Gastrointest Endosc 1991; 37:383-393.
7. Hawes RH, Cotton PB, Vallon AG. Follow-up 6 to 11 years after duodenoscopic sphincterotomy for stones in patients with prior cholecystectomy. Gastroenterology 1990; 98:1008-1012.
8. PereiraLima JC, Jakobs R, Winter UH et al. Long-term results (7 to 10 years) of endoscopic papillotomy for choledocholithiasis. Multivariate analysis of prognostic factors for recurrence of biliary symptoms. Gastrointest Endosc 1998; 48:457-464.
9. Tanaka M, Takahata S, Konomi H et al. Long-term consequence of endoscopic sphincterotomy for bile duct stones. Gastrointest Endosc 1998; 48:46-59.

CHAPTER 20

Endoscopic Therapy of Benign Pancreatic Disease

Martin L. Freeman

Introduction

ERCP has been used to treat a variety of biliary disorders for more than 20 years. Initially, the role of ERCP in pancreatic disease was primarily diagnostic, with intervention limited to biliary therapeutics in order to treat acute gallstone pancreatitis. Increasingly, in order to treat pancreatic disorders, endoscopic therapeutic techniques have been applied to the pancreatic sphincters and ducts.[1] While ERCP with biliary therapeutics are performed in most larger hospitals, and many smaller community hospitals, pancreatic therapeutic ERCP is primarily performed at tertiary-referral hospitals. Pancreatic endotherapeutic techniques are generally more difficult and riskier than their biliary counterparts, and the outcomes less predictable.[2] Pancreatic therapeutic ERCP requires a sophisticated unit with a highly skilled endoscopist, experienced GI assistants, and a broad range of endoscopic accessories. Collaboration with advanced surgeons and interventional radiologists is a prerequisite.

Anatomy

The pancreas is a retroperitoneal organ which extends from the medial wall of the duodenum, along the posterior wall of the stomach, and extending laterally to near the spleen. The exocrine function of the pancreas includes secretion of digestive enzymes and bicarbonate-rich juice via a ductal system into the duodenum. In normal patients, the main pancreatic duct branches and drains via two separate orifices, with the majority of the flow through the duct of Wirsung and major papilla, with lesser or no flow through the duct of Santorini and minor papilla (Fig. 20.1a). In patients with pancreas divisum, the most common congenital anomaly of the pancreatic ductal system, the ventral and dorsal ducts never fuse, so that the majority of pancreatic juice must exit via the minor papilla (Fig. 20.1b). Many congenital and acquired pancreatic diseases involve functional or structural obstruction to or disruption of outflow of pancreatic secretions, creating an opportunity for endoscopic diagnosis and therapy.

Indications

- **Biliary therapeutics** (biliary sphincterotomy+/ biliary stent) to treat pancreatic disease
  - Acute or recurrent gallstone pancreatitis with impacted stone in bile duct; with severe acute pancreatitis
    - For prevention of recurrent choledocholithiasis
  - Acute or recurrent biliary pancreatitis; microlithiasis (documented biliary microcrystals or sludge); type III choledochal cyst (choledochocele)

Gastrointestinal Endoscopy, edited by Jacques Van Dam and Richard C. K. Wong.

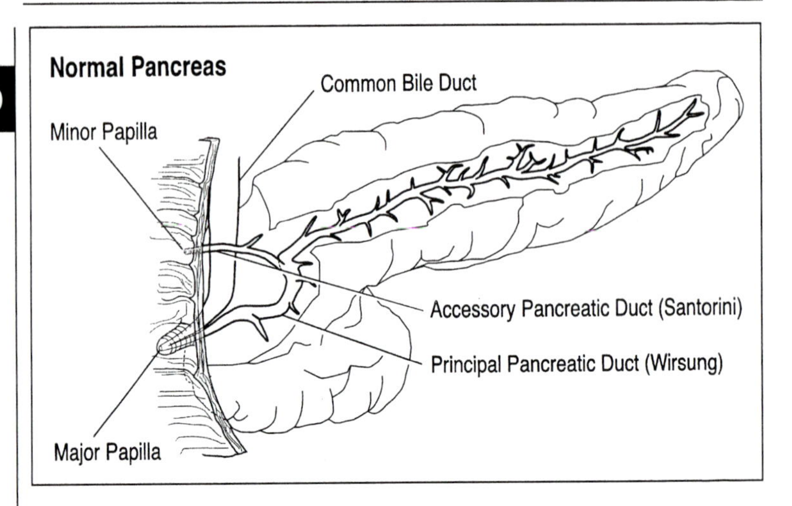

Fig. 20.1a. Anatomy of the pancreas: normal.

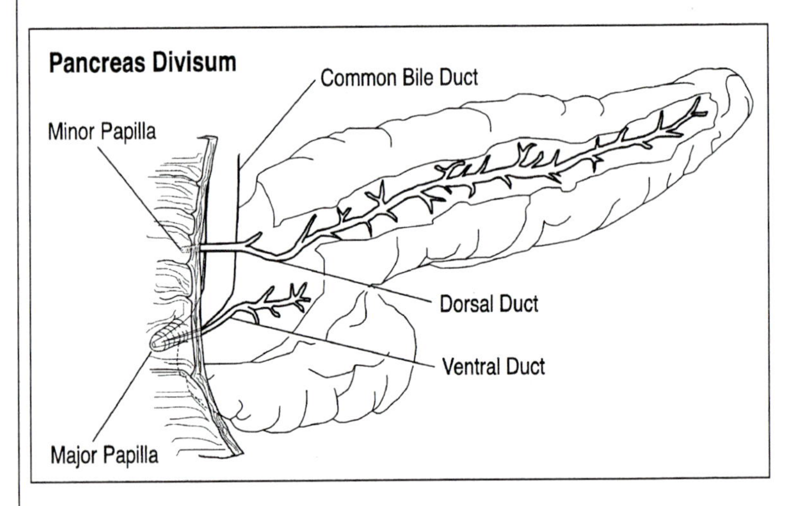

Fig. 20.1b. Anatomy of the pancreas: pancreas divisum.

(Fig. 20.2); miscellaneous: intradiverticular papilla, duodenal duplication cysts, etc.
- Acute recurrent pancreatitis due to sphincter of Oddi dysfunction / papillary stenosis (generally, pancreatic sphincterotomy is also required)
- Common bile duct strictures associated with chronic pancreatitis (biliary stent +/ biliary sphincterotomy)

• Diagnosis/treatment of unexplained acute recurrent pancreatitis
    - Biliary therapeutics for indications listed above

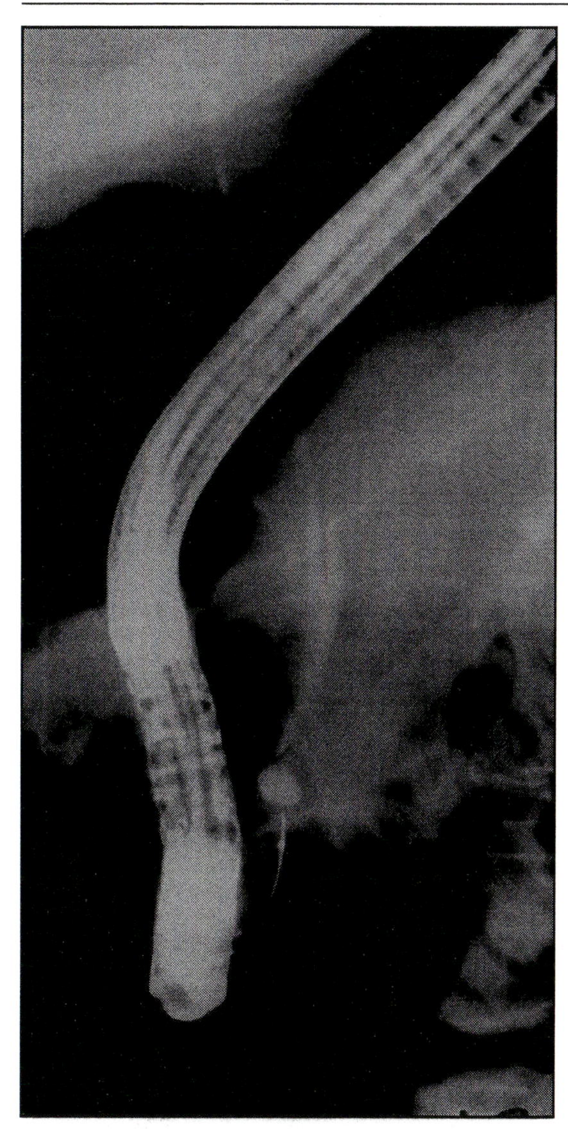

Fig. 20.2. Choledochocele (type III choledochal cyst) as cause for acute recurrent pancreatitis.

- Pancreatic therapy (sphincterotomy, stent, dilation, and / or stone removal) (major or minor papilla); pancreatic sphincter of Oddi dysfunction (major papilla); pancreas divisum (minor papilla therapy for "dominant dorsal duct" syndrome); pancreatic strictures (access via either papilla) chronic pancreatitis, post-inflammatory, idiopathic
  - pancreatic ductal stones (access via either papilla) chronic pancreatitis, post-inflammatory, idiopathic

- neoplasms (access via either papilla); pancreatic ductal adenocarcinoma; ampullary adenoma or carcinoma; mucinous ductal ectasia (intraductal papillary mucinous tumor); metastatic tumors to the pancreas; islet cell tumors of the pancreas
- Miscellaneous obstructive pancreatic anomalies (duodenal diverticulum, duodenal wall cyst, choledochal cyst, annular pancreas, etc.)

- Pancreatic ductal tissue sampling, diagnosis of suspected pancreatic neoplasm
- Palliation (stenting) of neoplastic pancreatic ductal stricture with recurrent pancreatitis, or pain
- Endoscopic therapy (stenting) of complete or partial pancreatic duct disruptions
- Diagnosis and treatment of unexplained abdominal pain suspected to be of pancreatic origin (including manometry and sphincterotomy of pancreatic and/or biliary sphincters)
- Diagnosis and palliative treatment of chronic pancreatitis (pancreatic sphincterotomy, stone extraction, stricture dilation/stenting)
- Treatment of pancreatic pseudocyst
  - Transpapillary drainage (pancreatic stent or nasopancreatic drain)
  - Transgastric or transduodenal cystenterostomy
- Pancreatic stenting as an aid to endoscopic access to bile duct

Contraindications (Relative)

- Unstable cardiopulmonary condition precluding prolonged procedure
- Severe uncorrectable coagulopathy or thrombocytopenia
- Obstructed access to the major or minor papilla because of oropharyngeal disease, esophageal, gastric or duodenal stenosis due to stricture or tumor, or inaccessibility to papillas because of long-limb surgical anastamoses or other reconstructive surgery
- Severe acute pancreatic necrosis

Equipment, Endoscopes, Devices and Accessories

Endotherapy of pancreatic disease requires a completely equipped, and highly sophisticated ERCP unit with high-resolution fluoroscopy, and staffed by GI assistants experienced at pancreatic endotherapeutic techniques.

- As for all endoscopy units, electrophysiologic monitoring must include including monitoring devices for EKG, blood pressure, and pulse oximetry, oxygen delivery systems, oral suction and emergency resuscitation equipment.
- Endoscopes should include a standard therapeutic channel (3.2-3.8 mm) and large therapeutic channel (4.2 mm) duodenoscope for adult pancreatic endotherapy. For pediatric or infant patients, smaller caliber or pediatric duodenoscopes are sometimes used.
- Devices and accessories
  - Cannulas ranging from 5 F standard-tip down to ultratapered 3 F tipped
  - Guidewires ranging from 0.035 down to 0.018 inch including hydrophilic guide wires
  - Papillotomes: standard 6 to 7 French diameter traction-type; small-caliber (5 French) wire-guided traction-type; needle-knife
  - Pancreatic stents (with multiple sideholes), in single pigtail and straight-type design: 5, 7, and 10 F (3,4 F optional), lengths ranging from 2 cm to 15 cm; double-pigtail 7 and 10 F stents for cystenterostomy

- Nasopancreatic drains, 5 and 7 F
- Stricture-dilating catheters from 3 F to 10 F diameter
- Dilating balloons, low profile, 4 mm, 6 mm, 8 mm, 10 mm
- Balloons and baskets (including standard and wire-guided) for stone and stent extraction
- Sclerotherapy needles, thermocoagulation (bipolar or heater probe) for hemostasis

Endoscopic Techniques

- Endoscopic biliary sphincterotomy, stricture dilation, stent insertion. See section on ERCP and biliary therapy.
- Endoscopic pancreatic sphincterotomy. Sphincterotomy may be performed on the major papilla or on the minor papilla, in order to facilitate access to the pancreatic duct, for dilation of strictures or placement of stents, to extract pancreatic duct stones, and to treat sphincter stenosis or dysfunction.
  - Techniques include: pull-type traction sphincterotomy (Fig. 20.3a, b and c); needle-knife over a pancreatic stent (Fig. 20.4); needle-knife access papillotomy
  - General Techniques for pancreatic sphincterotomy
    - Major papilla pancreatic sphincterotomy . Pancreatic sphincterotomy may be performed alone, after endoscopic biliary sphincterotomy, or as a combined procedure with sequential pancreatic then biliary sphincterotomy. In general, free cannulation of the pancreatic duct is obtained and a guide wire is placed through the cannula deep into the main pancreatic duct around the genu, or angulation between the head and body of the pancreas. With traction sphincterotomy, a Pull-type sphincterotome is passed over the guide wire until a small amount of cutting wire is in contact with tissue. Using short increments of electrocautery, the sphincter is divided in a step-wise fashion by 12 mm increments up to a usual maximum of 10-15 mm, with the size of the incision depending on the anatomic landmarks and the intended effect. Another common method is prior placement of a pancreatic stent followed by needle knife incision along the course of the stent (Fig. 20.4). Finally, access papillotomy can rarely be performed on the major papilla to gain entry into the main pancreatic duct. In this technique, a needle knife electrocautery wire is used to incrementally dissect open the major papilla until the pancreatic duct is found. This technique is particularly suitable if there is an impacted pancreatic duct stone, but is seldom used.
    - Minor papilla pancreatic sphincterotomy (Fig. 20.3 a, b and c). Minor papilla pancreatic sphincterotomy can be performed by the same three methods described for major papilla pancreatic sphincterotomy. Needle knife access papillotomy may occasionally be used to access the minor papilla, primarily for impacted pancreatic duct stones, or pancreas divisum, where an orifice cannot be identified despite simulation of pancreatic juice outflow by administration of intravenous secretin, and especially when a "Santorinicele," or cystically dilated intraduodenal terminus of the dorsal duct is identified.
- Pancreatic stent insertion (Fig. 20.5a and b). As for biliary stents, placement of a pancreatic stent requires free cannulation of the desired papilla and duct with

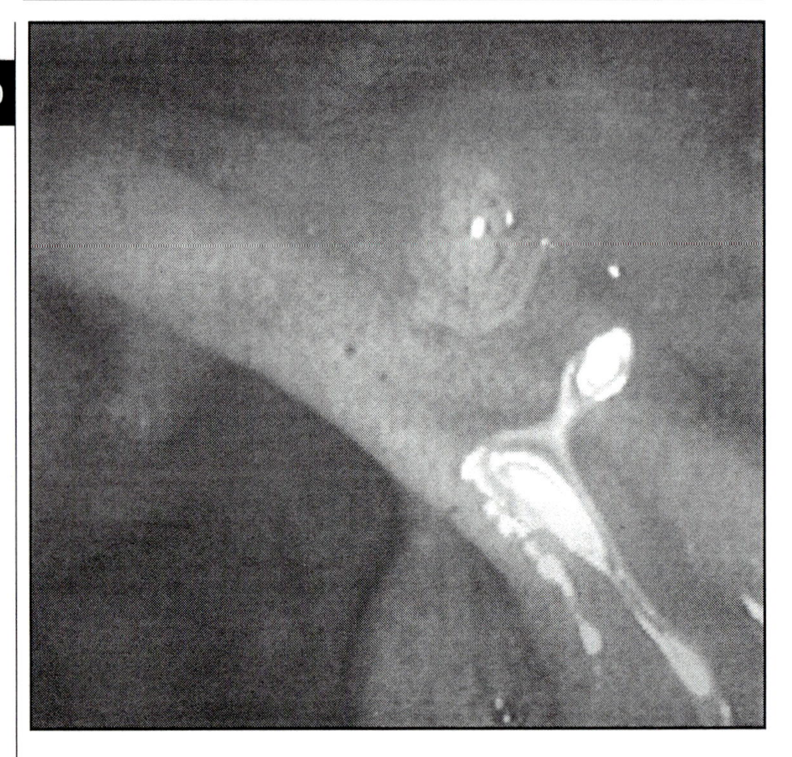

Fig. 20.3a. Standard traction-type pancreatic sphincterotomy of the minor papilla—intact minor papilla.

deep insertion of a guide wire, usually at least to the midbody. The stent is advanced over the guide wire with a pusher tube until it enters the duct and the outermost flaps are visible. With large bore stents such as 10 F the stent must be passed over an inner guide catheter. To prevent inward migration into the duct, most pancreatic stents utilize either double external flaps or a one-half to three-quarters pigtail on the duodenal end. Internal ends of pancreatic stents usually have either one or no flaps.

- **Pancreatic stricture dilation**. Pancreatic strictures are particularly common in patients with chronic pancreatitis, and dilation of these strictures can be extremely difficult. Stricture dilation requires prior passage of a guide wire through the stricture. For very small or tight strictures, graduated tapered dilators are sequentially passed over the guidewire, in increasing diameters from 3 F up to 7 to 10 F Occasionally, a corkscrew-type stent extractor can be used to bore a path through an otherwise undilatable stricture. Once the lumen of the stricture is at least 5 F diameter, a hydrostatic balloon dilator can be passed across the stricture and inflated to 10 to 12 atm. Stents are generally inserted after stricture dilation to maintain stricture patency and pancreatic drainage (Fig. 20.5a and b).

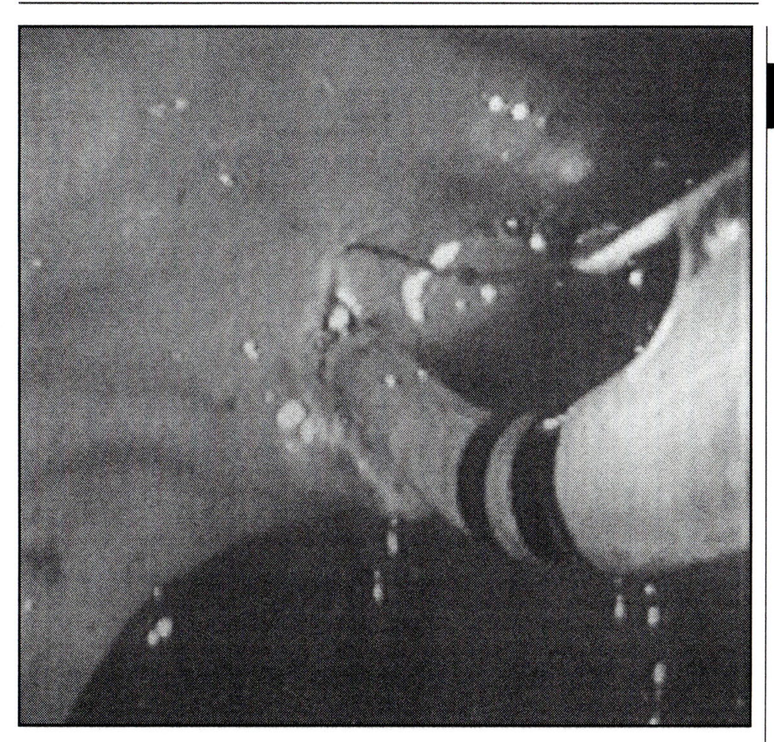

Fig. 20.3b. Standard traction-type pancreatic sphincterotomy of the minor papilla—sphincterotomy in progress.

- **Pancreatic stone extraction** (Fig. 20.6a-20.6c). Pancreatic stones usually occur in the setting of severe chronic pancreatitis. They may occur anywhere in the ductal system, from the main duct to side branches. Main pancreatic duct stones often cause secondary obstruction and may cause pain or flares in pancreatitis. Removal of pancreatic duct stones is more difficult than bile duct stones because they are hard and are frequently trapped above strictures. Successful removal of pancreatic duct stones generally requires a pancreatic sphincterotomy, basket or balloon extraction, and often the use of adjunctive fracturing techniques such as extracorporeal shock wave lithotripsy, or intraductal mechanical or electrohydraulic laser lithotripsy. Multiple procedures are often necessary to achieve total stone clearance in the main pancreatic duct.
- Pancreatic duct tissue sampling
  - **Brush cytology**. For suspected malignant lesions, a cytology brush may be passed over a previously place guide wire.
  - **Biopsy**. Pancreatic ductal biopsy is usually performed using a small angle-jaw biopsy forceps, which is usually passed along side a guide wire deeply seated in the pancreatic duct. Prior performance of the pancreatic sphincterotomy facilitates ductal access.

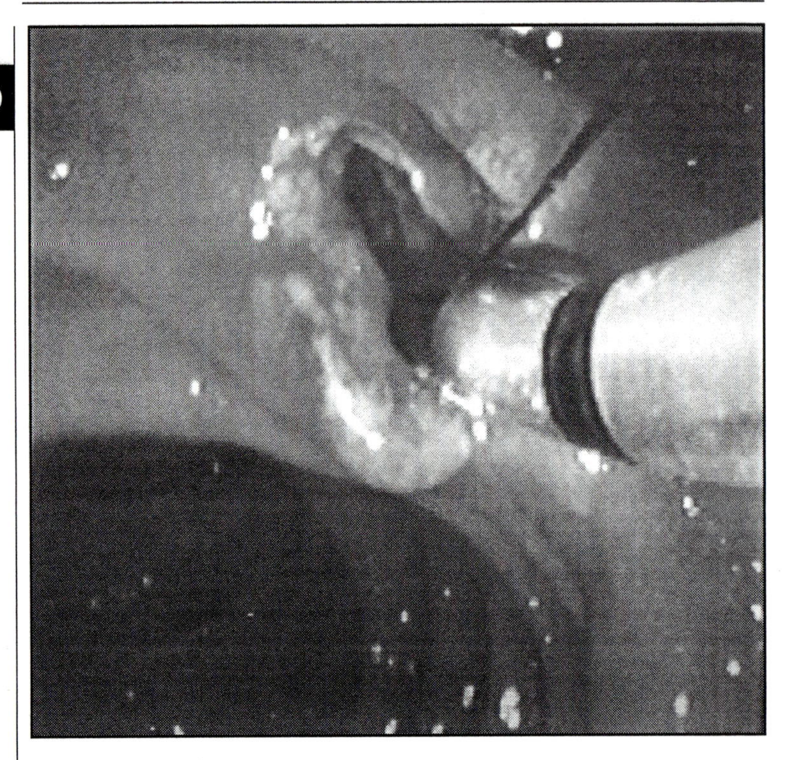

Fig. 20.3c. Standard traction-type pancreatic sphincterotomy of the minor papilla—completed sphincterotomy.

- **Fine needle aspiration**. Pancreatic strictures and/or suspected pancreatic tumors can be sampled by passage of a hollow needle aspirating catheter into the duct. The needle is extended into the area of the stricture and vigorous suction applied using a syringe.

- **Pancreatoscopy**. A "baby" scope may be inserted deep into the pancreatic duct through a therapeutic ERCP scope. Indications for pancreatoscopy include: 1) determining the extent of intraductal papillary mucinous tumors, 2) direct visualization of strictures for diagnosis and directed tissue sampling, 3) fragmentation of pancreatic stones under direct vision utilizing electrohydraulic lithotripsy or modified laser devices. Baby scopes currently available are approximately 9 F (3 mm) in diameter, but prototype scope as small as 1 to 2 mm using microchips are being developed. Pancreatoscopy requires two operators, one to drive the ERCP scope and a second operator to drive the baby scope. Generally, pancreatoscopy can only be performed in the setting of a dilated pancreatic duct and requires a pancreatic sphincterotomy. The pancreatic duct is freely cannulated and a guide wire passed deep into the duct. Then the baby scope is inserted over the guide wire by a second operator who maneuvers the tip of the scope into the pancreatic sphincter and duct.

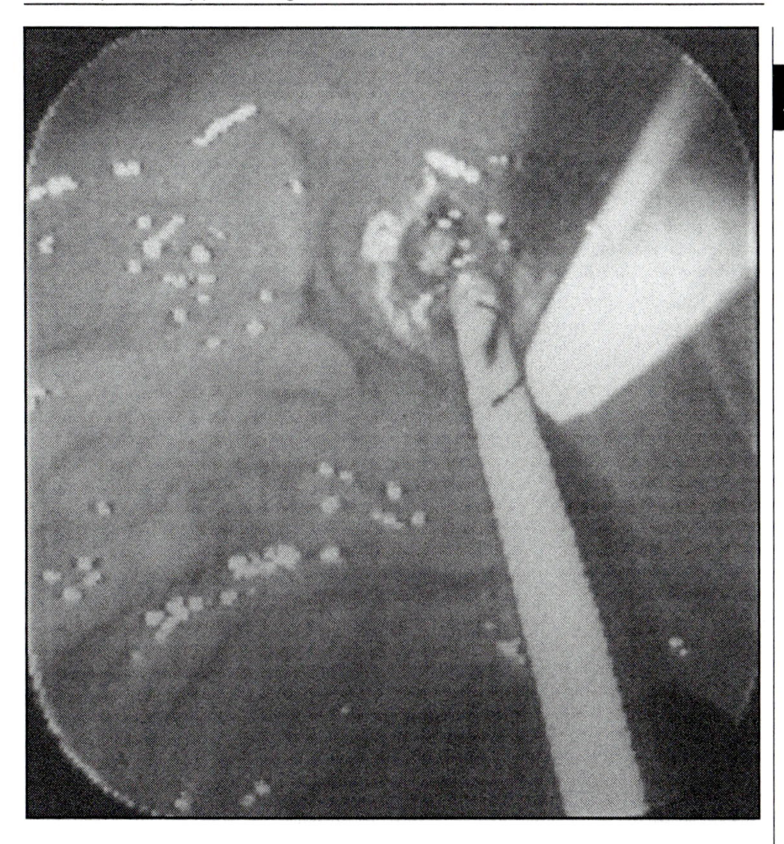

Fig. 20.4. Pancreatic sphincterotomy via needle-knife incision over pancreatic stent (major papilla).

- Pancreaticobiliary sphincter of Oddi manometry. In general, investigation of suspected functional disorders of the sphincter of Oddi as a cause for recurrent acute pancreatitis or unexplained abdominal pain requires manometry of the pancreatic as well as the biliary sphincter. Sphincter of Oddi manometry involves measurement of pressures in the sphincters and ducts via a triple lumen catheter, most of which currently utilize two perfusing ports to measure pressures and the third port to aspirate infused water or saline in order to prevent excessive pressure build up inside the pancreatic duct.[3] The manometry catheter is deeply inserted into the pancreatic duct, usually freehand, but occasionally over a previously placed guidewire, and is then withdrawn in 2 mm increments through the sphincter zone while pressures are recorded. Techniques and interpretation of biliary and pancreatic manometry are similar except that duration and frequency of pull-through's in the pancreatic duct is usually limited to reduce risk of iatrogenic pancreatitis. In addition, continuous aspiration through the third lumen is often performed during pancreatic manometry to prevent buildup of excessive pressure in the duct.

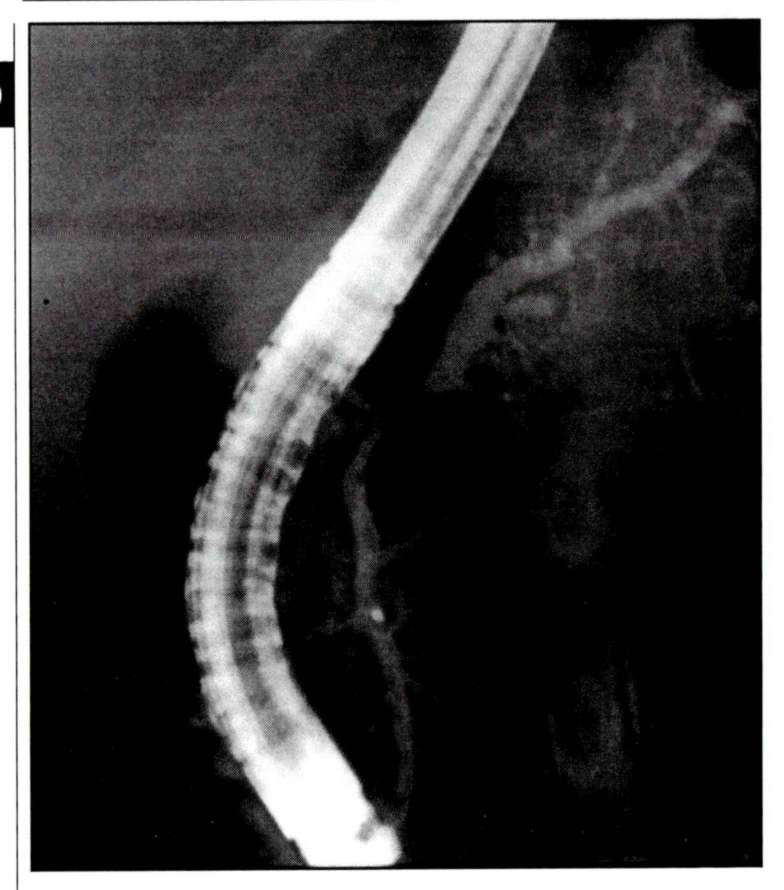

Fig. 20.5a. Pancreatic stent insertion—pancreatic stricture with upstream dilation in patient with chronic pancreatitis.

- Pseudocyst drainage. **Pseudocysts** are liquefied collections resulting from acute or chronic pancreatitis,

and may or may not communicate with the main pancreatic duct. Endoscopic techniques for pseudocyst drainage include transpapillary drainage through the pancreatic duct, and transgastric or transduodenal cystenterostomy.

- **Transpapillary drainage**. In this technique, a stent may be placed through the major or minor papilla into the pancreatic duct beyond the level of a duct disruption which communicates with a cyst, or in the case of large cysts, directly into the cyst cavity. Occasionally, transpapillary stents are placed both into the main duct and into the cyst. If there is a downstream stricture or stone, a passage of a stent beyond the obstructing lesion will often result in resolution of the pseudocyst. Smaller pseudocysts without downstream ductal obstruction will often resolve with transpapillary stenting alone.

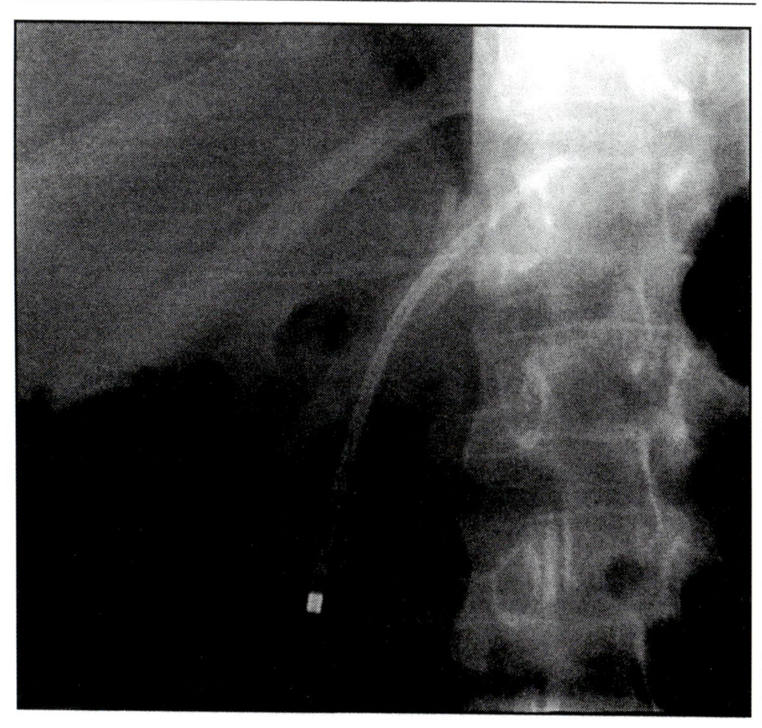

Fig. 20.5b. Pancreatic stent insertion—stent inserted through stricture.

- **Cystenterostomy** (Fig. 20.7a-20.7d). Larger pseudocysts which are contiguous with the duodenal or gastric wall may be endoscopically drained directly through the stomach or duodenal wall. This technique is preferred for larger pseudocysts and for those which contain debris. Eligibility for endoscopic cystenterostomy is determined by CT scanning and/or endoscopic ultrasound demonstrating close apposition of the pseudocyst wall to the bowel wall. Identification of the puncture site is made by visualizing a bulge in the gastric or duodenal wall corresponding with the location of the pseudocyst (Fig. 20.7a). In cases where the location of the pseudocyst is not obvious, endoscopic ultrasound may be useful to identify the optimal site and perform direct puncture under real-time endosonographic guidance. In addition, endoscopic ultrasound may be useful to evaluate for underlying blood vessels or varices in the planned puncture tract. Through a therapeutic duodenoscope, a Howell aspiration needle with an extended length of 8 mm may be passed through the wall of the stomach, and contrast injected to localize the pseudocyst under fluoroscopy guidance. Cyst puncture is performed with a needle-knife sphincterotome, by making a very small puncture hole through the wall of the stomach or duodenum into the cyst. Once it is deeply inserted, the catheter sheath is left in place, the needle-knife core is removed, and fluid can be aspirated for analysis. Then, contrast can be

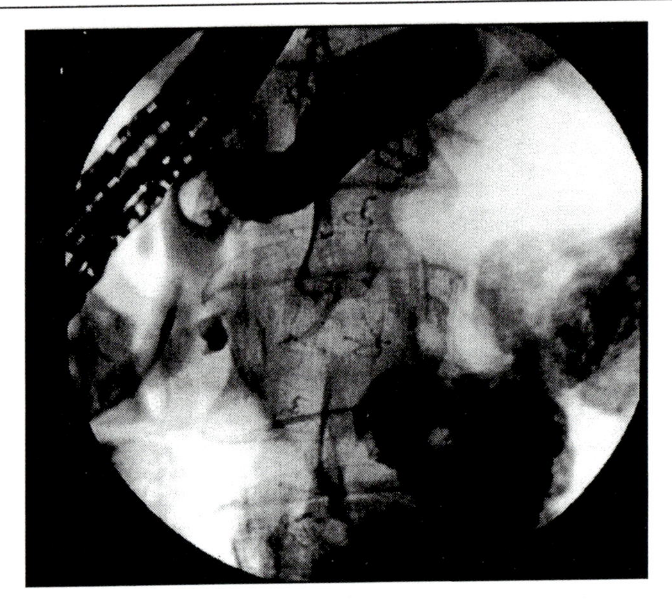

Fig. 20.6a. Pancreatic stone extraction (multiple main duct stones in patient with hereditary pancreatitis)—multiple large stones in dilated main duct.

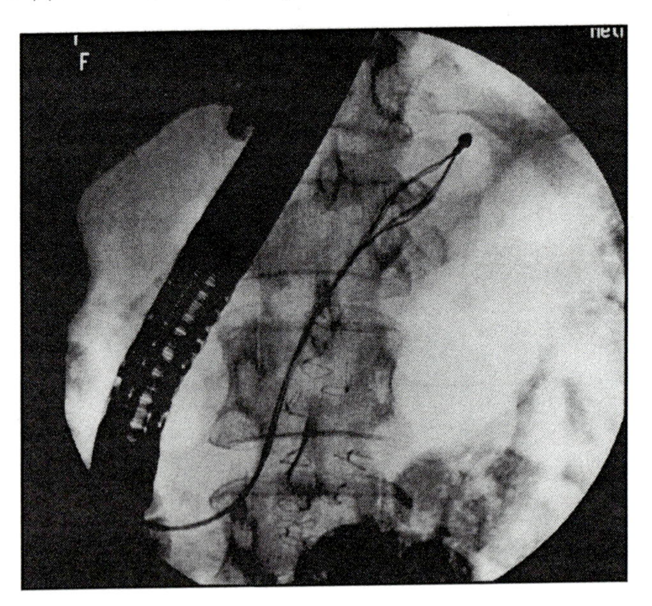

Fig. 20.6b. Pancreatic stone extraction (multiple main duct stones in patient with hereditary pancreatitis)—after pancreatic sphincterotomy, basket used to extract stones.

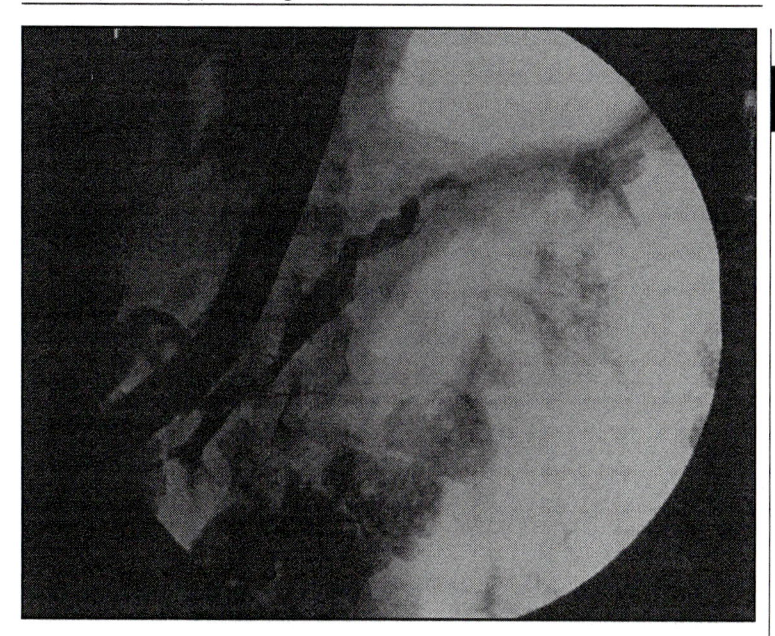

Fig. 20.6c. Pancreatic stone extraction (multiple main duct stones in patient with hereditary pancreatitis)—total clearance and decompression of main duct.

injected to outline the pseudocyst cavity (Fig. 20.7b) and a guide wire passed into the pseudocyst cavity in the cavity to maintain access. The fistula tract is dilated using a 10 mm or greater diameter balloon (Fig. 20.7c). One or more double pigtail stents are inserted to maintain the tract patency (Fig. 20.7d). When there is debris in the fluid, or in case of postnecrotic pseudocysts, a nasocystic drain can be placed through the nose and cystenterostomy into the pseudocyst cavity to allow saline lavage of the cyst contents.

- **Pancreatic stenting** to aid endoscopic access to bile duct
  - Prior to needle-knife "precut" biliary access papillotomy
  - Before or after biliary sphincterotomy to reduce risk of ERCP-induced pancreatitis

Outcomes

- Biliary therapeutics (i.e., biliary sphincterotomy)
  - **Acute gallstone pancreatitis.** Endoscopic biliary sphincterotomy to remove distal bile duct stone in patients with acute gallstone pancreatitis has been shown in a number of studies to reduce the severity of pancreatitis, especially in severe cases and to reduce other associated complications such as cholangitis.[4] Outcomes are best if sphincterotomy is performed within the first 24 hours, and its effectiveness decreases with delay of the procedure up to several days. In addition, biliary sphincterotomy may prevent recurrence

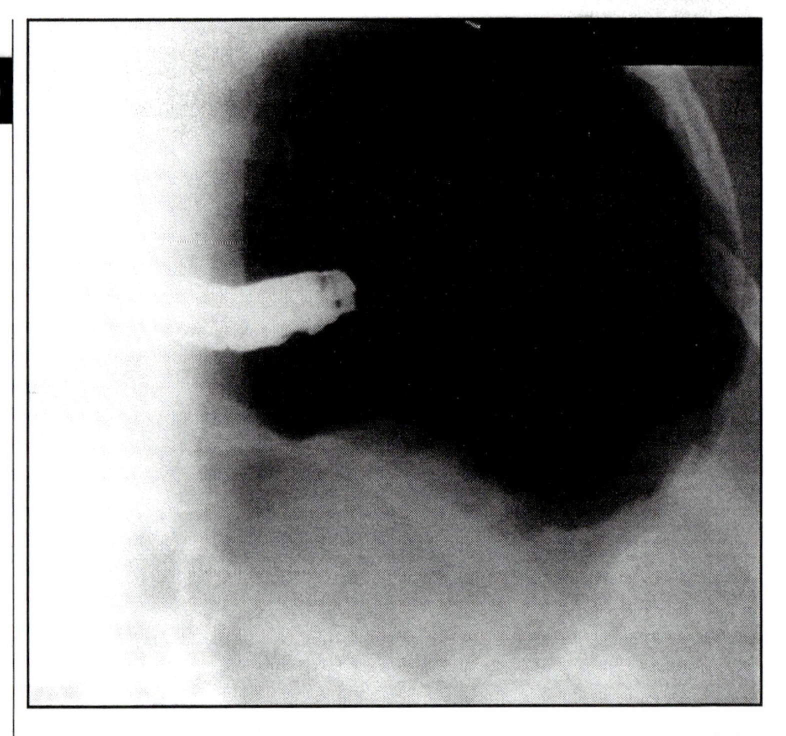

Fig. 20.7a. Transgastric pseudocyst drainage (cystenterostomy)—pseudocyst bulge into gastric outline visible on fluoroscopy.

of gallstone pancreatitis when the offending stone has already passed, but the patient is not in good condition to undergo surgery. However, in some cases, significant damage to the pancreas has already occurred by the time ERCP is performed, and some patients may go on to develop severe, necrotizing or complicated pancreatitis despite spontaneous passage or endoscopic removal of the offending stone. Patients with presumed gallstone pancreatitis but few clinical or laboratory indicators of impacted common bile duct stones are unlikely to benefit from preoperative ERCP and are best served with a cholecystectomy when the pancreatitis has resolved.

- **Microlithiasis**. Patients determined to have microlithiasis by ultrasonography or bile analysis may benefit from endoscopic biliary sphincterotomy, with decreased or no relapses of pancreatitis.[5]
- Miscellaneous other causes of biliary pancreatitis such as type III choledochal cyst (choledochocele) (Fig. 20.2) are usually definitively treated by biliary sphincterotomy.
- For treatment of acute recurrent pancreatitis secondary to sphincter of Oddi dysfunction, biliary sphincterotomy alone is often ineffective, and fairly risky.[6] A pancreatic sphincterotomy with a temporary pancreatic stent is usually also required.[7]

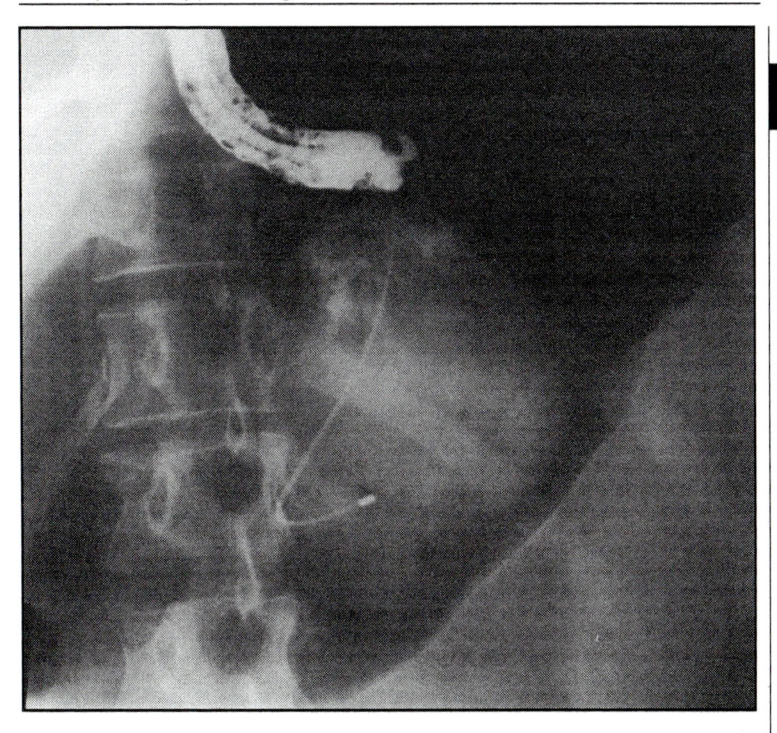

Fig. 20.7b. Transgastric pseudocyst drainage (cystenterostomy)—after cyst puncture, contrast is injected to outline cavity and place guidewire.

- For **common bile duct strictures** due to chronic pancreatitis, biliary stenting and/or dilation may provide temporary palliation, but surgical biliary-digestive anastamosis is often required for definitive management.[8]

• Diagnosis and treatment of unexplained **recurrent acute pancreatitis**.
  - Documented biliary causes of pancreatitis respond well to biliary sphincterotomy alone, but empirical biliary sphincterotomy without identification of a specific problem is risky[6] and often ineffective. Advanced diagnostic and therapeutic management of unexplained acute recurrent pancreatitis requires a number of specialized diagnostic techniques including sphincter of Oddi manometry, minor papilla cannulation, pancreatic fluid analysis and cytology, and the ability to perform directed pancreatic therapeutics including major and minor papilla pancreatic sphincterotomy and stenting. Diagnostic evaluation using advanced techniques for unexplained acute pancreatitis reveals a cause in at least 75% of cases, most of which can also be treated at the same session by advanced endoscopic methods.[5]
  - **Pancreatic sphincterotomy** and/or **stenting of major or minor papilla**
    - Endoscopic therapy for sphincter of Oddi dysfunction as a cause for acute recurrent pancreatitis is generally effective if pancreatic sphincterotomy is performed in addition to biliary sphincterotomy, but there are few data on long term outcomes.[6]

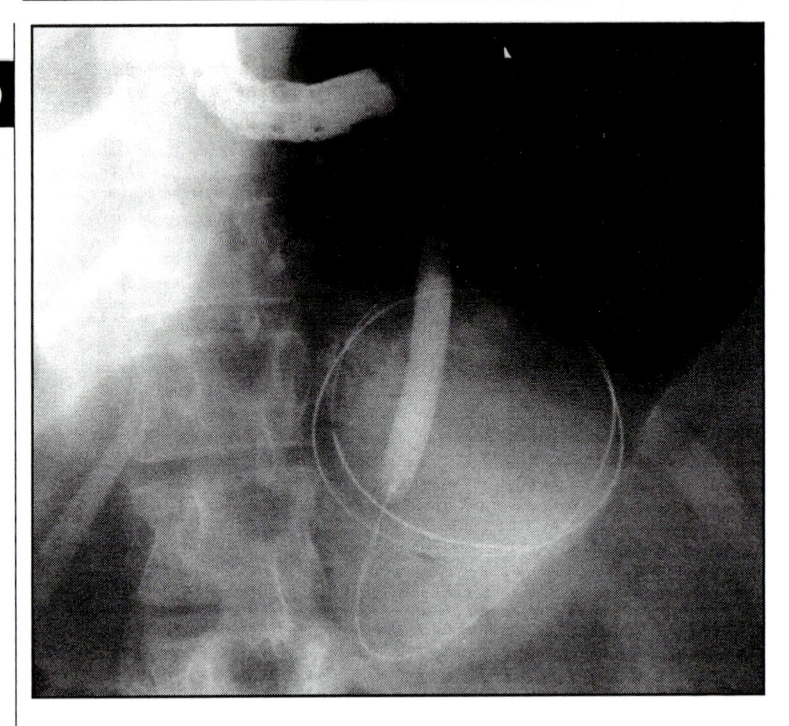

Fig. 20.7c. Transgastric pseudocyst drainage (cystenterostomy)—cystgastrostomy tract is dilated with 10 mm balloon.

- For pancreas divisum, minor papilla sphincterotomy results in good responses in about 80% of patients with acute recurrent pancreatitis only, 40% of patients with chronic pancreatitis in the dorsal duct, but only 20% of patients with no clinical evidence of pancreatitis but chronic abdominal pain only. [9]
- Long term response to endoscopic therapy of pancreatic ductal strictures is variable. Sometimes, repeated dilation and long term stenting will result in permanent resolution of strictures and prevention of recurrent pancreatitis, but stenoses can be refractory to endoscopic treatment, and require surgical management.[10]
- Removal of obstructing main pancreatic duct stones is usually effective in preventing recurrent attacks of acuteonchronic pancreatitis and often (60-90%) effective in reducing or relieving pain in patients with chronic pancreatitis.[11]
- Acute or smoldering pancreatitis as a result of malignant strictures of the pancreatic duct often responds dramatically to placement of pancreatic stents through the stricture. Palliation of pain without evidence of ongoing pancreatitis is variable and often disappointing.

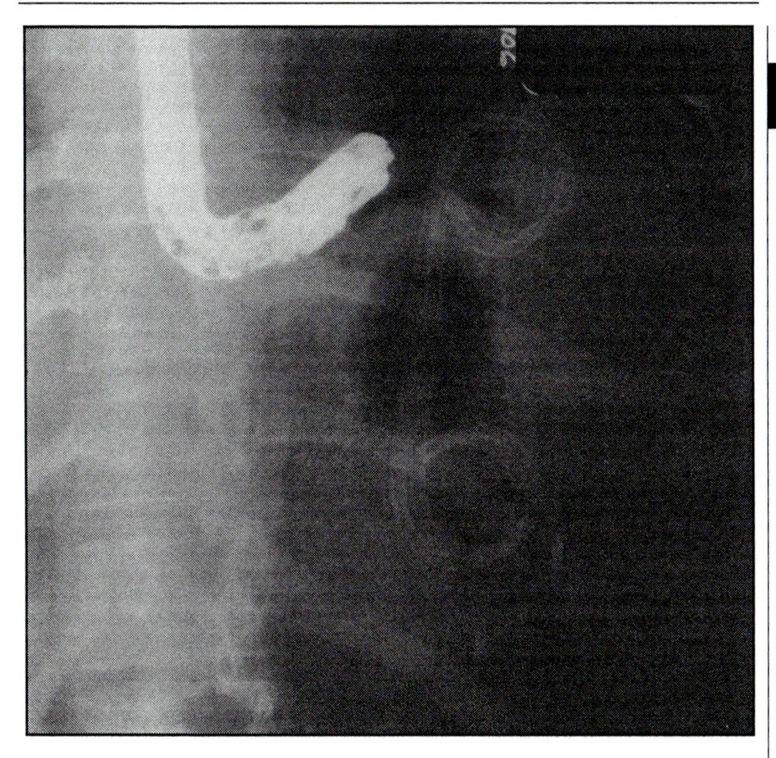

Fig. 20.7d. Transgastric pseudocyst drainage (cystenterostomy)—two double pigtail 10 F stents are placed to maintain tract patency.

- Yield of tissue sampling for pancreatic neoplasms is variable with sensitivities up to 75% for combined methodologies and specificities of up to 100%.[12]
- Malignant strictures of the pancreatic duct.
- Temporary transpapillary pancreatic duct stenting is usually successful in closing partial pancreatic duct disruptions.[13] Complete transections of the pancreas and duct as sometimes seen in blunt traumatic injuries to the abdomen do not respond to endoscopic therapy and require surgical resection of the discontinuous portion of the pancreas.
- Diagnosis and treatment of unexplained pain suspected to be of pancreatic origin but without concrete evidence of pancreatic disease is controversial, as few data are available, but can occasionally be dramatic. For patients with pain only, and abnormal pancreatic manometry, pancreatic sphincterotomy is occasionally performed during the index procedure in combination with biliary sphincterotomy, but more often at a later date if biliary sphincterotomy fails to relieve unexplained pain.[7] Results of endoscopic therapy for pancreas divisum are discussed above.

- Endoscopic therapy for palliation of pain in chronic pancreatitis has variable results. Most favorable responses generally occur with extraction of obstructing distal pancreatic duct stones. Outcome is less certain when there are residual strictures and/or advanced parenchymal disease.[11] Pain relief is variable for stenting and or dilation of dominant strictures in chronic pancreatitis. Strictures may occasionally permanently resolve with endoscopic therapy but long term stenting is often required.[10] It is unclear whether favorable short-term response to pancreatic stenting predicts a favorable response to surgical decompressive therapies such as a Peustow procedure (lateral pancreaticojejunostomy).
- Endoscopic pseudocyst drainage results in 70 to 100% resolution with recurrence rates of 6-20%, depending in part on whether any underlying pancreatic ductal disease was simultaneously treated. In general, endoscopic drainage of pseudocysts has similar success, complication, and relapse rates to surgical drainage. Combinations of therapeutic modalities including endoscopic transpapillary and transluminal drainage, or endoscopic and percutaneous radiological drainage or surgical drainage are sometimes required.[14]
- In order to reduce risk of iatrogenic pancreatitis, pancreatic stenting is increasingly used in advanced endoscopy centers to aid in accessing the bile duct during precut sphincterotomy or in conjunction with biliary sphincterotomy in high-risk patients such as those with sphincter of Oddi dysfunction. There are few published data.

Complications

- Pancreatitis
- Hemorrhage
- Pancreatic and biliary sepsis
- Perforation
  - retroperitoneal from sphincterotomy
  - ductal from guidewires
- Stent related
  - occlusion
  - migration into or out of the duct
  - ulceration/obstruction by migrated stent
- General cardiopulmonary and other complications related to ERCP, sedation and analgesia
- **Pancreatitis** is the most common complication of ERCP and is usually defined by a clinical picture of pancreatitis, that is new or worsened abdominal pain, associated with an amylase level at least three times normal at more that 24 hours after the procedure, and requiring unplanned admission of an outpatient or extension of a planned admission for more than one day after the procedure. Pancreatitis is graded as mild if admission is prolonged for 2-3 days, moderate if it is prolonged for 4-10 days and severe if hospitalization of more than 10 days or any of the following occur: pancreatic necrosis, pseudocyst, or intervention requiring percutaneous or surgical drainage. Treatment requires fasting, supportive care, and rarely interventional drainage procedures.[6] Pancreatitis after ERCP can occasionally be severe and life-threatening with multiorgan failure and even death. In general, endoscopic pancreatic therapy is relatively safer in patients with advanced chronic pancreatitis, but riskier (10-30%) in patients

without chronic pancreatitis, particularly those with sphincter of Oddi dysfunction, or patients with pancreas divisum undergoing dorsal duct therapy.[2]

- patients should be observed for 2-8 hours after ERCP for development of abdominal pain, or nausea and vomiting
- If pain or other abdominal symptoms occur, patients should be kept NPO, blood should be drawn for amylase, lipase, and CBC and admission to hospital is advised for close observation

• **Hemorrhage** occurs in under 12% of patients after sphincterotomy.[2,6] Presentation is typical for any upper GI hemorrhage, but may present up to 10 days after sphincterotomy.[6] Diagnosis and supportive care is typical of any upper GI hemorrhage. Diagnosis and therapy are performed at endoscopy using a side-viewing duodenoscope. Very rarely, angiographic embolization or surgery are required. If patients develop melena or hematemesis after ERCP

- Vital signs should be checked, including orthostatic blood pressure changes
- Blood should be drawn for hgb
- IV access established
- Emergency endoscopy should be considered
- Admission to the hospital if any evidence of significant past or ongoing bleeding

• **Pancreatic infection** or sepsis occasionally occurs with occlusion of a pancreatic stent or with inadequate endoscopic drainage of an obstructed pancreatic duct. **Biliary sepsis** may occur with pancreatic therapy without concomitant biliary drainage or with incompletely drained biliary obstruction. Treatment usually involves antibiotics and exchange of the occluded or malfunctioning stent. If patients develop fever or chills after ERCP

- Vital signs should be checked
- Blood should be drawn for CBC, liver and pancreatic chemistries
- Abdominal flat and upright series to evaluate position of stents and possible perforation.
- Admission to hospital

• **Perforation** occurs in under 1% of pancreatic or biliary sphincterotomies and is recognized by presence of abdominal pain, often with fever and/or leukocytosis, and occasionally crepitus in the neck or chest. Distinctions between perforation and pancreatitis can be difficult, and both complications can occur simultaneously, but presence of abdominal pain despite a normal serum amylase or lipase should raise suspicion of perforation. Radiographic confirmation of perforation is made by CT scan demonstrating retroperitoneal air or fluid, and rarely manifests as free intraperitoneal air on abdominal x-rays. If recognized early, perforations can often be managed conservatively with antibiotics, nasogastric suction, and ideally nasopancreatic and/or nasobiliary drainage, but surgical drainage and/or diversion is sometimes required. If patients develop significant abdominal pain after ERCP with sphincterotomy

- Patient should be kept NPO
- Blood drawn for amylase, lipase, CBC
- If amylase is normal, or there is other suspicion of perforation
  - Abdominal flat and upright. Normal study does not rule out perforation
  - Spiral CT scan obtained, including examination of "lung windows" for detection of retroperitoneal air or fluid

- **Stent-related complications** are common with pancreatic endotherapy.[2] Most pancreatic stents occlude within 6-12 weeks. Occlusion usually presents as pain or infection. Inward migration of stents requires potentially challenging endoscopic or even surgical removal and can be prevented by using stents with duodenal pigtails or not more than one internal flange. Outward migration, or passage of pancreatic stents rarely causes problems other than recurrence of pancreatic ductal obstruction and usually pass through the GI tract without problem. Indwelling pancreatic stents may cause chronic ductal injury with stricturing. Stent-related injury occurs primarily in the absence of chronic pancreatitis. As a result, pancreatic stents should be left in for the minimum possible duration (17 days) in patients without chronic pancreatitis. Pancreatic sphincter restenosis may occur at variable intervals after pancreatic sphincterotomy and can usually be retreated with a repeat sphincterotomy; however, deeper extension of pancreatic sphincterotomy may result in scarring and stricturing as pancreatic tissue is cut.
- Like any endoscopic procedure, ERCP may result in esophageal perforation, allergic medication reactions, and cardiopulmonary or other general medical morbidity, especially in elderly or frail patients. Diagnosis and management depend on the type of complication.

Selected References

1. Sherman S. Endoscopic therapy of pancreatic disease. Gastrointest Endosc Clin N Am 1998; 8:12-72.
2. Freeman ML, DB Nelson, DiSario JA et al. Outcomes of pancreatic therapeutic ERCP as compared with biliary therapeutic and diagnostic ERCP: A prospective multisite study (abstract) Gastrointest Endosc 1998; 47:AB136.
3. Sherman S, Troiano FP, Hawes RH et al. Sphincter of Oddi manometry: Decreased risk of clinical pancreatitis with use of a modified aspirating catheter. Gastrointest Endosc 1990; 36:462-466.
4. Soetikno RM, CarrLocke DL. Endoscopic management of acute gallstone pancreatitis. Gastrointest Endosc Clin N Am 1998; 8:1-12.
5. Tarnasky PR, Hawes RH. Endoscopic dignosis and therapy of unexplained (idiopathic) acute pancreatitis. Gastrointest Endosc Clin N Am 1998; 8:13-38.
6. Freeman ML, Nelson DB, Sherman S et al. Complications of endoscopic biliary sphincterotomy. N Engl J Med 1996; 335:909-918.
7. Kuo WH, Pasricha PJ, Kalloo AN. The role of sphincter of Oddi manometry in the diagnosis and therapy of pancreatic disease. Gastrointest Endosc Clin N Am 1998; 8:79-86.
8. Ng C, Huibregtse T. The role of endoscopic therrapy in chronic pancreatitisinduced common bile duct strictures. Gastrointest Endosc Clin N Am 1998; 8:181-194.
9. Lehman GA, Sherman S. Diagnosis and therapy of pancreas divisum. Gastrointest Endosc Clin N Am 1998; 8:39-54.
10. Binmoeller KF, Rathod VD, Soehendra N. Endoscopic therapy of pancreatic strictures. Gastrointest Endosc Clin N Am 1998; 8:125-142.
11. Deviere J, Delhaye M, Cremer M. Pancreatic duct stones management. Gastrointest Endosc Clin N Am 1998; 8:163-180.
12. Lee JG, Leung JW. Tissue sampling at ERCP in suspected pancreatic cancer. Gastrointest Endosc Clin N Am 1998; 8:221-236.
13. Kozarek RA. Endoscopic therapy of complete and partial pancreatic duct disruptions. Gastrointest Endosc Clin N Am 1998; 8:39-54.
14. Howell DA, Elton E, Parsons WG. Endoscopic management of pseudocysts of the pancreas. Gastrointest Endosc Clin N Am 1998; 8:143-162.

CHAPTER 21

ERCP in Malignant Disease

William R. Brugge

Introduction

Endoscopic retrograde cholangiopancreatography (ERCP) is commonly performed for malignant diseases of the pancreatic-biliary tree. The procedure may be performed for diagnosis or therapy. Although the techniques employed are often complex and difficult, the basis of the procedure is relatively simple. This chapter will introduce the physician in training to the anatomic basis of the procedure, indications, equipment, technique, and outcome, including complications.

Anatomy of the UGI Tract

- Access to the ampulla of Vater is through the pylorus and duodenum in the intact UGI tract.
- After a partial gastrectomy (Billroth II), access to the ampulla of Vater is difficult and consists of retrograde passage of the duodenoscope through the efferent limb of the gastrojejunostomy.
- After an antrectomy (Billroth I), the access is through the duodenogastrostomy and is relatively simple.
- After a Whipple resection, access to the ampulla is made in a retrograde fashion through the afferent limb of the gastrojejunostomy.

Anatomy of the Ampulla of Vater

- There is a great deal of variation in the appearance of the major ampulla of Vater. The ampulla may appear as a small button of tissue, a large protuberance, or a thickened fold.
- The ampulla can usually be located by finding a large transverse fold in the second part of the duodenum. It is located along the medial wall of the duodenum and 24 cm distal to the duodenal bulb.
- There may be a diverticulum adjacent to or encompassing the ampulla.
- The ampulla may be covered by a duodenal fold and visible only after pulling the fold off the ampulla.
- Bile staining of the mucosa can be a clue to locating the ampulla.
- The minor ampulla is usually 23 cm proximal to the major ampulla. The minor ampulla is smaller than the major ampulla and there is no bile staining. The minor ampulla is just distal the duodenal bulb, along the medial wall of the duodenum.
- In the absence of pancreas divisum, the major ampulla will contain the openings to the major pancreatic duct and the bile duct. In the vast majority of cases, there is one opening that houses both ducts. If there are separate openings, the bile duct is usually on the superior aspect of the ampulla.

Gastrointestinal Endoscopy, edited by Jacques Van Dam and Richard C. K. Wong.

Table 21.1. Indications for therapeutic ERCP

- Obstructive jaundice
- Acute cholangitis
- Stent occlusion
- Ampullary mass with bleeding and/or obstruction

- The minor ampulla has one opening and it is very small. Secretin stimulation may help in locating the opening since pancreatic juice can be seen flowing out of the opening.

Indications and Contraindications

- Therapeutic indications for ERCP
  - Obstructive jaundice. The presence of dilated bile ducts on CT/US is a strong predictor of biliary obstruction and is the most common indicator for ERCP. Cholestatic LFT's reflect the presence of biliary obstruction and often the first evidence of biliary disease.
  - Acute cholangitis usually occurs with evidence of biliary obstruction, often from a stone or stent.
    Occlusion of a biliary stent
  - Other Indications
    - Bleeding from the ampulla or tumor in the duodenal wall.
    - Chronic, uncomplicated obstruction of the pancreatic duct is not a frequent indication for ERCP.
  - Suspicion for an occult pancreatic-biliary malignancy
  - Need to obtain tissue for a histologic diagnosis of a pancreatic-biliary malignancy.
  - Differentiation between a pancreatic, biliary, and an ampullary malignancy.
  - Differentiation between chronic pancreatitis and cancer.
  - Diagnosis, biopsy and resection of ampullary tumors.
- Contraindications
  - Patients that are at high risk for hypoxia and hypotension.
  - Uncooperative patients.
  - Severe coagulopathy .
  - Gastric outlet obstruction.

Equipment

- Side viewing endoscopes are the mainstay of ERCP.
  - Video or fiber-optic imaging
  - Diagnostic or therapeutic channel size
- Accessories for cannulation and stenting
  - Papillotomes
    - 1, 2, or 3 lumens
    - short or long nose
  - Tapered catheters
    - plastic or metal tipped
    - long or short taper

Table 21.2. Diagnostic indications for ERCP

- Suspicion for occult pancreatic-biliary malignancy
- Need for tissue diagnosis of malignancy
- Differentiating between pancreatic, biliary, and ampullary tumors
- Differentiating between chronic pancreatitis and cancer
- Diagnosis of ampullary malignancy

Table 21.3. Contraindications for ERCP

- High risk for conscious sedation
- Severe coagulopathy
- Pyloric or duodenal obstruction

- Cannulation wires
  - with or without hydrophilic coating
  - stiff or flexible
- Stents
  - Biliary
    - plastic (8.5-11 Fr)
      - straight
      - pigtail
    - metal
      - coated or uncoated
  - Pancreatic
    - plastic (5-7 F)
- Drainage catheters
  - Pigtail
  - Straight

Technique

- Scope Placement
  - Passage through the pylorus and duodenum
  - Scope is straightened
  - Placement directly adjacent to the ampulla
- Cannulation
  - Duodenal motility is controlled with glucagon
  - Catheter is placed in superior aspect of ampulla for the bile duct and inferiorly for the pancreatic duct
  - Catheter or guide wire is gently advanced
  - Contrast is injected after cannulation is achieved.
- Sphincterotomy
  - Guide wire and catheter are placed in duct of choice
  - Tension and cautery are applied to the cutting wire
  - Needle-knife can be used in place of sphincterotome

- Stent Placement
  - Stent is placed in front of "pusher" catheter
  - Stent is pushed through the scope and into duct
  - Correct placement is confirmed fluoroscopically
  - "Pusher" catheter and wire are removed

Complications

- Immediate
  - Hypoxia, hypotension
  - Bleeding
  - Perforation
- Subacute
  - Cholangitis
  - Pancreatitis

CHAPTER 22

Endoscopic Ultrasound: Tumor Staging (Esophagus, Gastric, Rectal, Lung)

Manoop S. Bhutani

Introduction

Endoscopic ultrasonography or EUS is a technique where a high frequency (512 megahertz) transducer is built into the tip of an endoscope. When passed into the gastrointestinal tract (perorally or perrectally) these instruments provide high resolution images of the GI tract and adjoining structures within 45 cm of the gastrointestinal wall.

Relevant Anatomy

- By EUS five layers of the GI wall are seen (Fig. 22.1) with a total thickness of 34 mm.[1]
- These five layers correlate with histology of the gastrointestinal wall as shown in Table 22.1
- During staging of esophageal, gastric, lung and rectal tumors by EUS a number of anatomical structures are visualized in the mediastinum, upper abdomen and perirectum.[2]
  - Mediastinum(Fig. 22.2)
  - Esophageal wall
  - Aorta
  - Pulmonary Artery
  - Azygous vein
  - Heart
  - Trachea
  - Spine
  - Thoracic duct
  - Upper Abdomen(Fig. 22.3)
  - Gastric wall
  - Aorta
  - Celiac artery
  - Splenic artery
  - Common hepatic artery
  - Pancreas
  - Splenic vein
  - Portal vein – splenic vein confluence
  - Liver
  - Gall bladder
  - Rectum

Gastrointestinal Endoscopy, edited by Jacques Van Dam and Richard C. K. Wong.

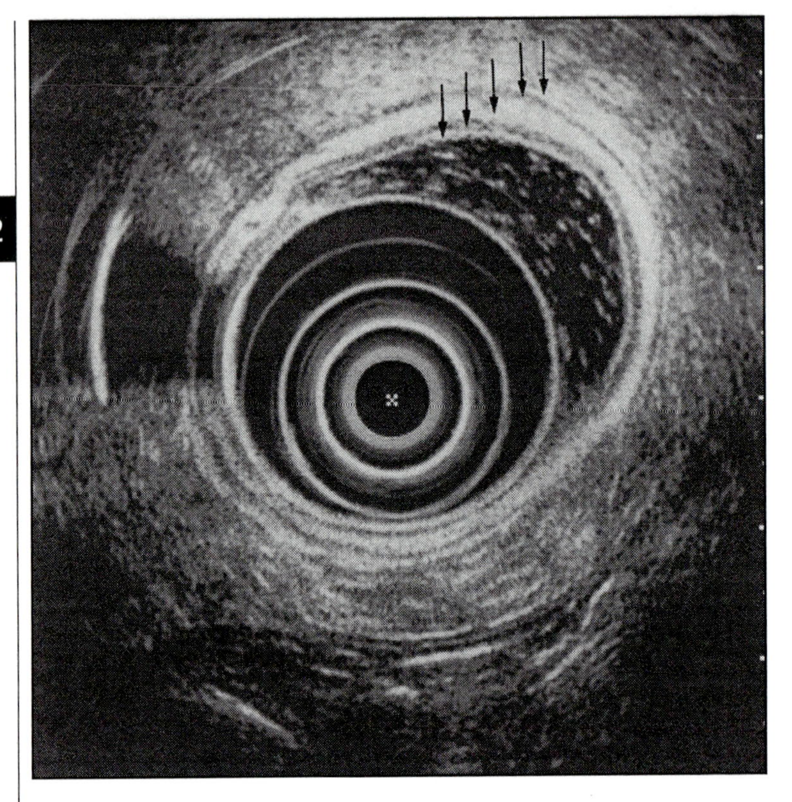

Figure 22.1. Five echo layers of the gastrointestinal wall (arrows) as seen by endoscopic ultrasound.

Table 22.1. Histologic correlates of EUS images of gastric wall

EUS Layer	Histologic Correlate
1st EUS layer (hyperechoic)	Mucosa (superficial)
2nd EUS layer (hypoechoic)	Mucosa (deep)
3rd EUS layer (hyperechoic)	Submucosa
4th EUS layer (hypoechoic)	Muscularis Propria
5th EUS layer (hyperechoic)	Serosa or Adventitia

- Rectal wall
- Urinary bladder
- Prostate and seminal vesicles in men
- Vagina in women

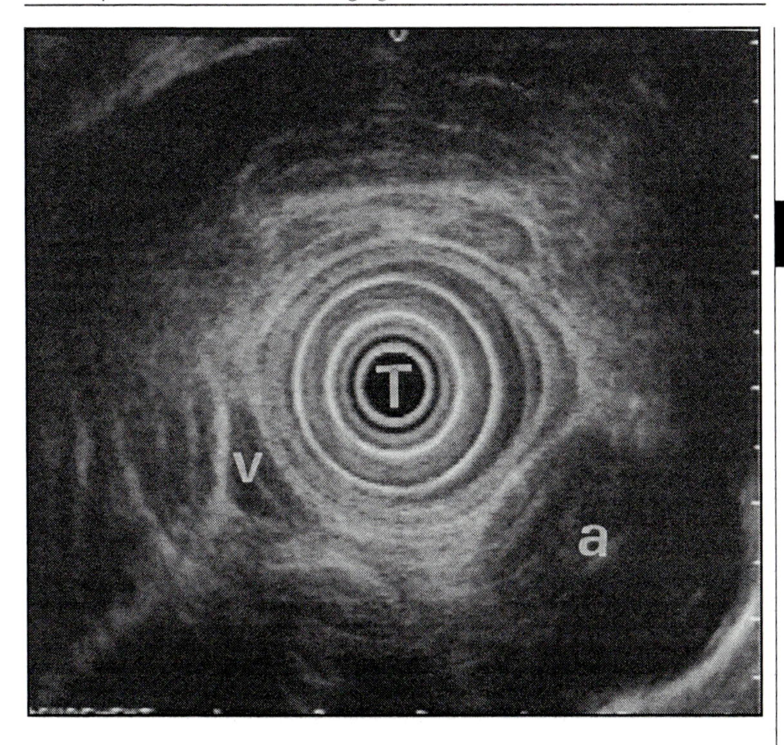

Figure 22.2. Mediastinal anatomy with the endoscopic ultrasound transducer (T) in the esophagus. a = Aorta, V = Azygous Vein.

Indications for EUS Tumor Staging

- Assessment of the extent of tumor invasion within the GI wall and to adjacent organs
- T stage by the TNM classification
- Assessment of lymph nodes for invasion by malignant cells; N stage by TNM classification

Contraindications for EUS Tumor Staging

- When the patient already has distant metastases (e.g., liver) by other imaging methods (e.g., CT scan or transabdominal US)
- When the results from EUS staging are not going to change the therapeutic strategy
- When the patient has comorbid factors (e.g., severe COPD, severe or unstable cardiac disease, etc.) such that he or she is at a high risk for conscious sedation and endoscopy.

Equipment for Endoscopic Ultrasound

- Radial GFUM20 or GFUM30 Echoendoscope (Olympus Corp.) (Fig. 22.5)
  - Scan Range: 360° (at rightangle to the long axis of the scope

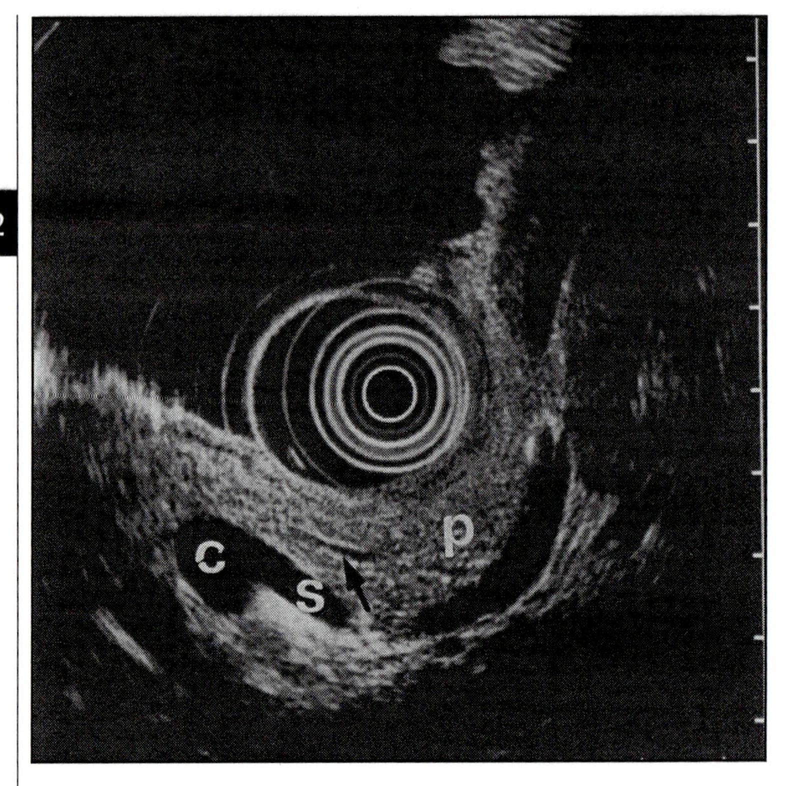

Figure 22.3. Retroperitoneal anatomy during transgastric endoscopic ultrasound imaging. p = pancreas, S = splenic vein, C = portal vein-splenic vein confluence, Arrow = main pancreatic duct.

  - Frequency: 7.5 MegaHertz-12 MegaHertz
  - Endoscopic View: 45° oblique
  - A newer generation of video radial echoendoscopes is available now that provides a much improved video endoscopic image.
- FG32UA, FG36UX, FG38UX linear array echoendoscopes (Pentax Corp.)(Fig. 22.6)
  - Scan range: 100° convex linear (parallel to the long axis)
  - Frequency: 5 megahertz-7.5 megahertz
  - Endoscopic view: 60° oblique
- GF – UM30P Mechanical Sector Scanning echoendoscope (Olympus Corp.)
  - Scan range: 250° mechanical sector (parallel to the long axis.
  - Frequency: 7.5 megahertz
  - Endoscopic view: 45° oblique
- GFUC 30P linear array echoendoscope (Olympus Corp.)
  - Scan range: 180° convex linear (parallel to the long axis)

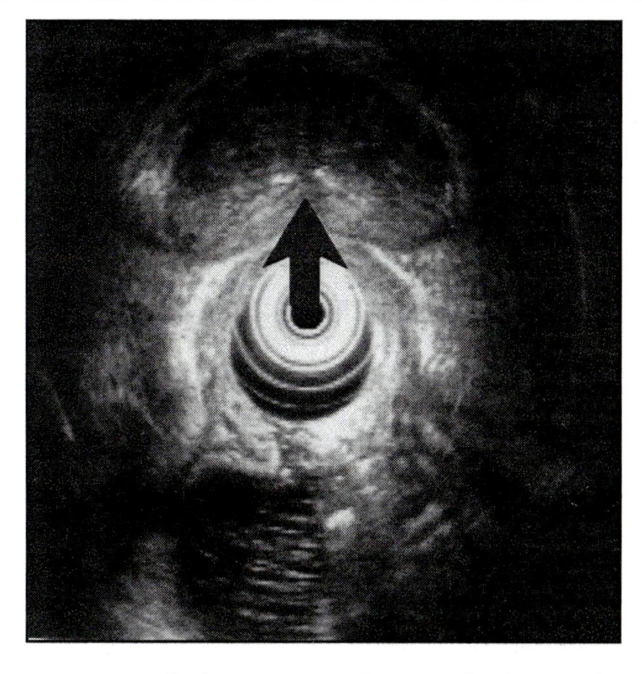

Figure 22.4a. Prostate gland (arrow) as seen during rectal endoscopic ultrasound.

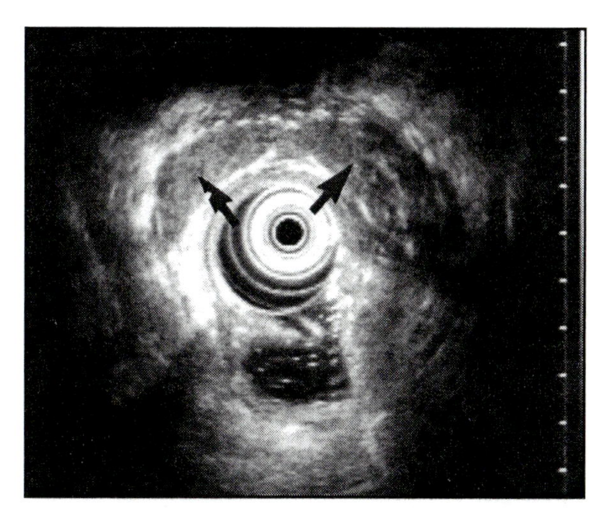

Figure 22.4b. Left and right seminal vesicles as seen during rectal endosonography.

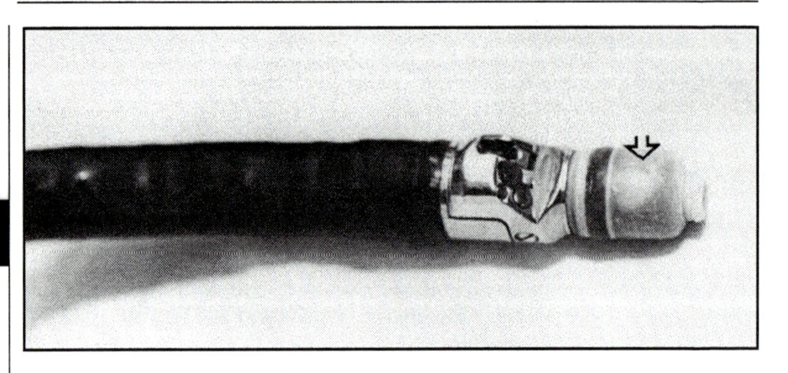

Figure 22.5. A radial echoendoscope with a rotating transducer that scans in a 360° fashion at right angles to the long axis of the instrument. arrow = transducer.

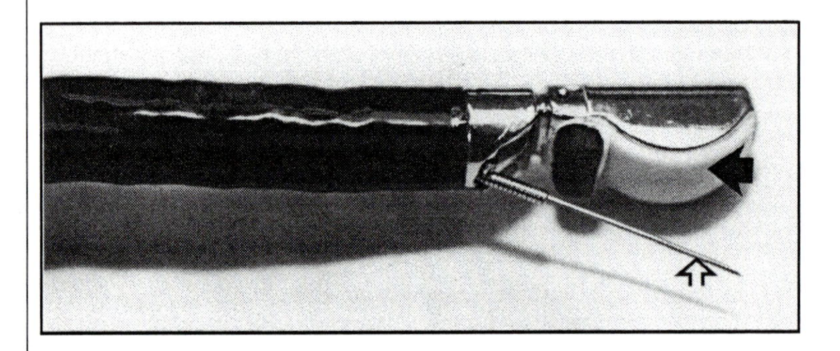

Figure 22.6. A linear array endoscopic ultrasound instrument that scans parallel to the long axis of the echoendoscope allowing ultrasound guided biopsies. Closed arrow = transducer, open arrow = needle for ultrasound guided biopsy.

  - Frequency: 7.5 megahertz
  - Endoscopic view: 45° oblique
- Endoscopic ultrasound guided biopsy can be performed by all four instruments of the above but is easiest with #2 and #4, fair with #3 and the most difficult with #1.
- Endoscopic ultrasound through the scope miniprobes (Fig. 22.7)
  - These slender catheter type probes can be passed through the biopsy channel of standard endoscopes for performing high frequency ultrasound
  - They are applied under direct vision to the region of interest
  - The probes operate at frequencies ranging from 1220 megahertz
  - Ultrasound guided biopsy is not possible
  - These are manufactured by Olympus Corp., Fujinon Corp., and Boston Scientific /Microvasive Corp.

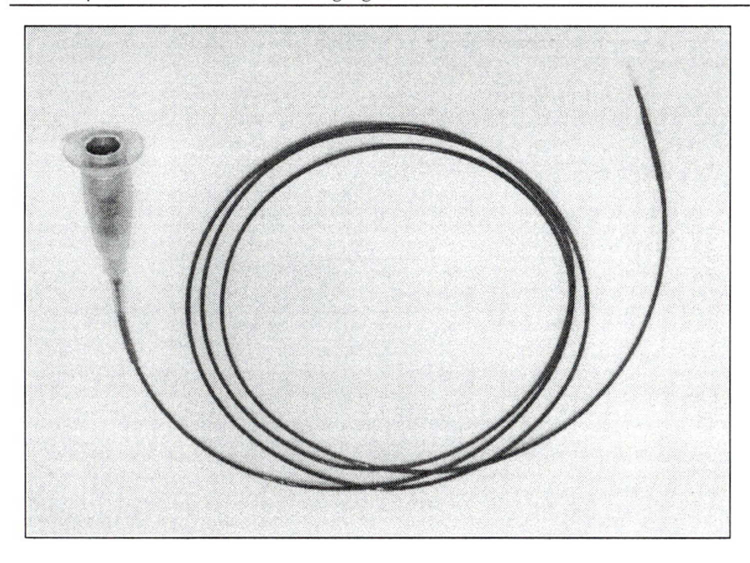

Figure 22.7. A catheter endoscopic ultrasound probe that can be passed through the biopsy channel of a standard endoscope into the gastrointestinal lumen.

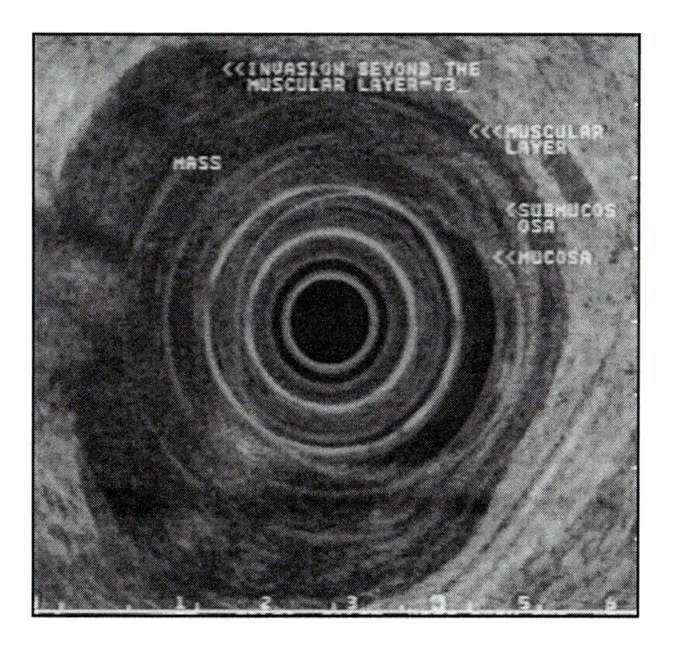

Figure 22.8. Rectal carcinoma imaged by endosonography as a hypoechoic mass involving the rectal wall. The five echo layer structure is somewhat preserved on the right half this image while complete destruction of these layers is seen on the left half.

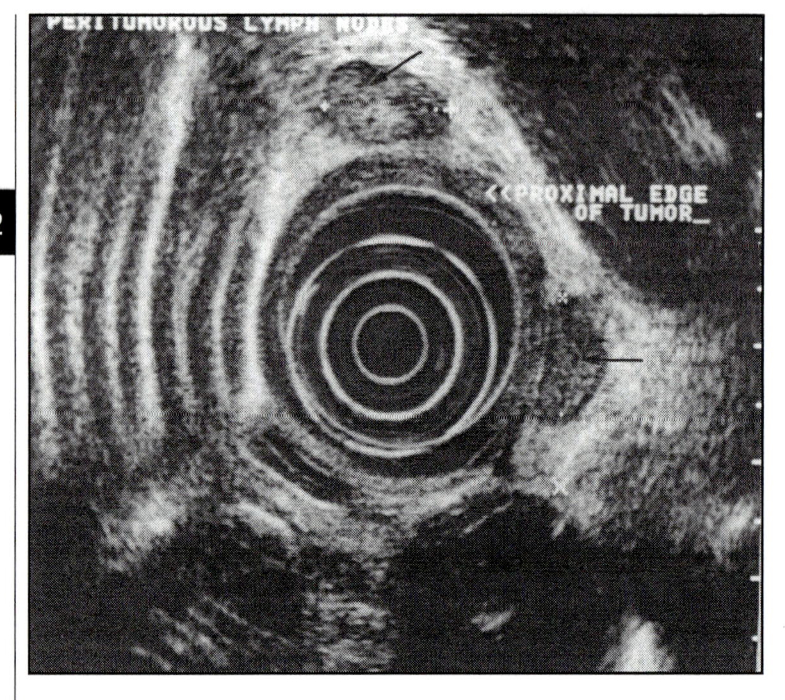

Figure 22.9. Oval, more than 1 cm in diameter lymph nodes in the mediastinum(arrows) imaged by endoscopic ultrasound in a patient with esophageal cancer.

General Technique of Endoscopic Ultrasound for Tumor Staging

- Informed consent.
- Conscious sedation.
- Standard upper endoscopy for esophageal or gastric cancer.
- Standard flexible sigmoidoscopy for rectal cancer.
- Echoendoscope passed into the esophagus, stomach or rectum.
- Balloon surrounding the echoendoscope is filled with water to displace air.
- Ultrasound transducer is turned on.
- If ultrasound transducer images are unsatisfactory, the GI lumen is filled with water to displace air and provide an acoustic interface.
- Primary tumors (esophageal, gastric and rectal cancer) are generally seen as hypoechoic masses arising from the gastrointestinal wall. The number of EUS layers of the GI wall invaded with the tumor is assessed as well as invasion of adjacent organs.[2] (Fig. 22.8)
- Lymph nodes (Fig. 22.9) are seen as oval, round or triangular structures outside the gastrointestinal wall in the mediastinum (in esophageal and lung cancer), stomach or the rectum.

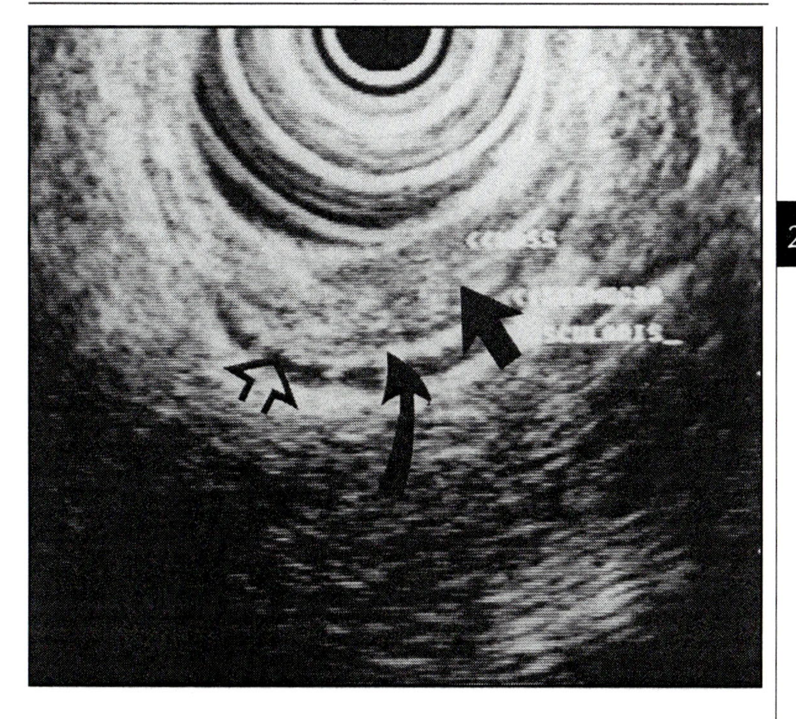

Figure 22.10. A T1 esophageal cancer (closed straight arrow) involving only the mucosa (1st and 2nd EUS layers.) Note the intact submucosa (curved arrow) and the intact muscularis propria (open arrow) under the lesion.

- An analysis of the echo features of lymph nodes is made by EUS:
  - Size
  - Echogenicity
  - Shape
  - Margins
- Some echo-features of lymph nodes may suggest malignancy.[3]
  - Size > 1 cm
  - Hypoechogenicity
  - Round shape
  - Distinct margins
- For definitive diagnosis of malignant invasion of lymph nodes trans-gastrointestinal EUS guided lymph node fine needle aspiration is performed.[4]

Results and Outcome

- Based on the above stated principles esophageal, gastric and rectal carcinomas are staged for T (tumor) and N (lymph node) stages according to the TNM classification.[5]

  These tumors are staged by EUS as follows according to the international TNM classification:

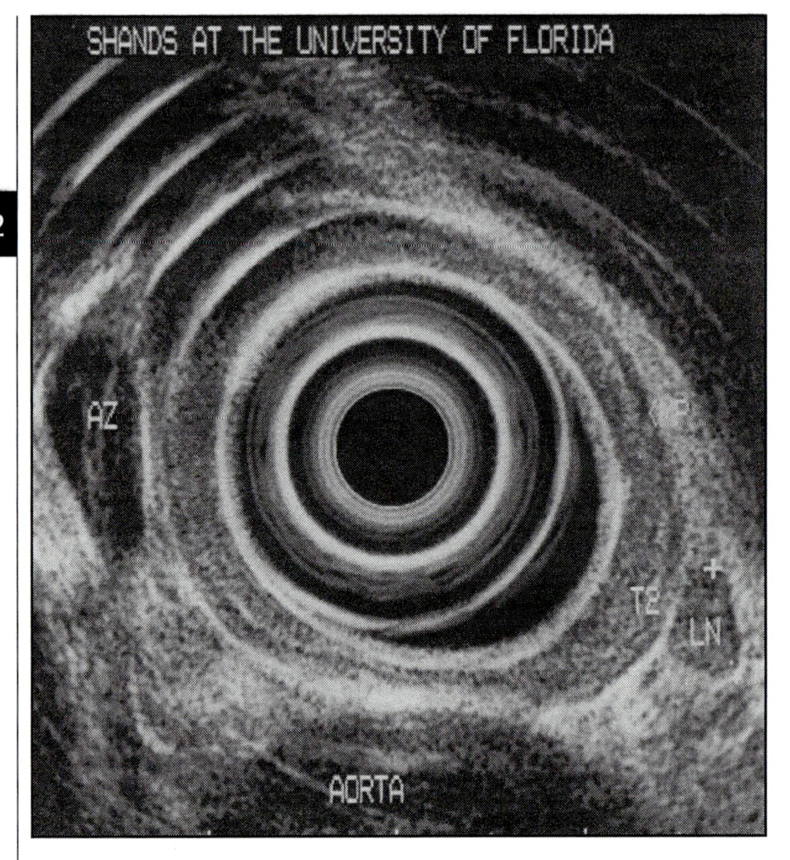

Figure 22.11. A T2 esophageal cancer that by EUS is penetrating into the muscularis propria (MP) with inability to define MP under the mass. AZ = azygous vein, LN = peritumorous lymph node.

- Esophageal Carcinoma
  - T : – Primary tumor stage
  - T1:—Tumor invades mucosa or submucosa (Fig. 22.10)
  - T2:—Tumor invades muscularis propria but does not penetrate beyond the muscularis propria (Fig. 22.11)
  - T3:—Tumor invades adventitia
  - T4:—Tumor invades adjacent structures (e.g., aorta) (Fig. 22.12)
  - N:— Regional lymph node staging
  - No:— No regional lymph node metastasis
  - N1:— Regional lymph node metastasis
- Gastric carcinoma
  - T:— Primary tumor stage
  - T1:— Tumor invades mucosa or submucosa but not the muscularis propria

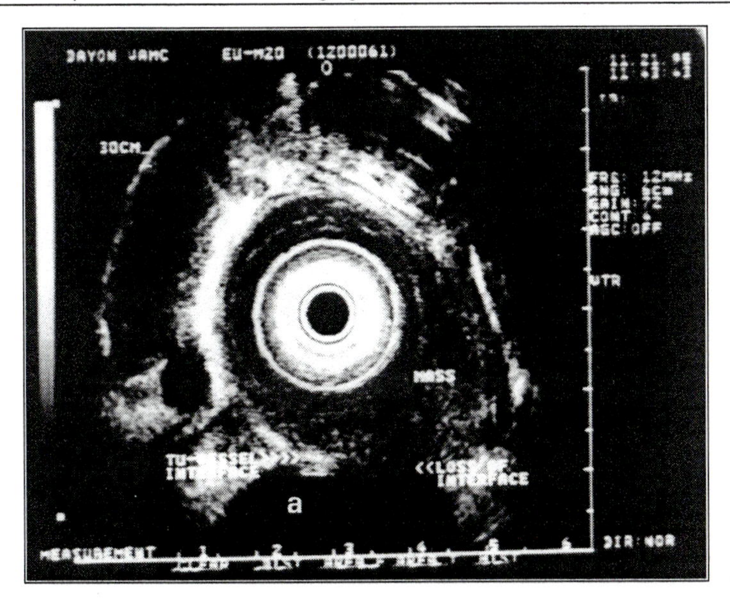

Figure 22.12. A T4 esophageal carcinoma by EUS appearing as a very hypoechoic mass with loss of interface between the aorta (a) and the mass, consistent with aortic invasion.

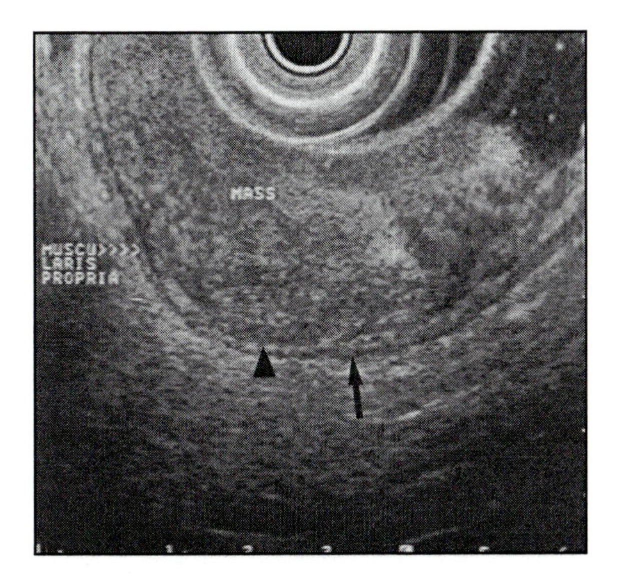

Figure 22.13. A T1 rectal carcinoma by EUS limited to the mucosa and the submucosa (arrow head). Note the intact muscularis propria under the malignant rectal mass (arrow).

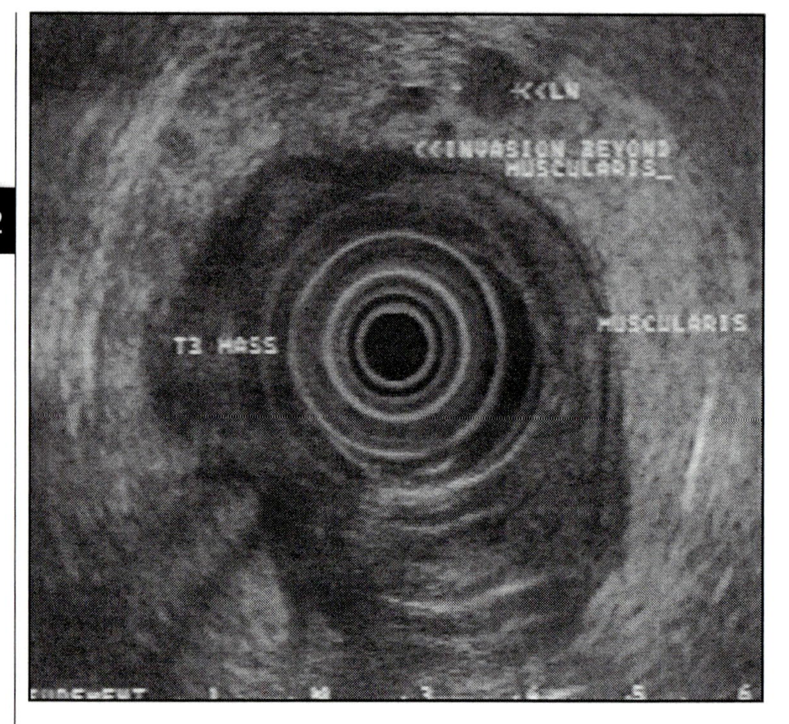

Figure 22.14. A T3 rectal cancer by EUS appearing as a hypoechoic lesion invading and penetrating through all the layers of the rectal wall into perirectal fat. LN = peritumorous lymph node.

  - T2:— Tumor invades muscularis propria or subserosa
  - T3:— Tumor penetrates serosa (visceral peritoneum) without invasion of adjacent structures
  - T4:— Tumor invades adjacent structures
  - N:— Regional Lymph Node Staging
  - N0:—No regional lymph node metastasis
  - N1:—Metastasis in 1 to 6 regional lymph nodes
  - N2:—Metastasis in 7 to 15 regional lymph nodes
  - N3:—Metastasis in >15 regional lymph nodes
- Rectal Carcinoma
  - T:— Primary tumor stage
  - T1:—Tumor invades up to the submucosa with no invasion of muscularis propria (Fig. 22.13)
  - T2:—Tumor invades muscularis propria

    T3:—Tumor invades through the muscularis propria into subserosa or perirectal tissues (Fig. 22.14)
  - T4:— Tumor directly invades other organs and structures
  - N:— Regional lymph node staging
  - NO:—No regional lymph node metastasis

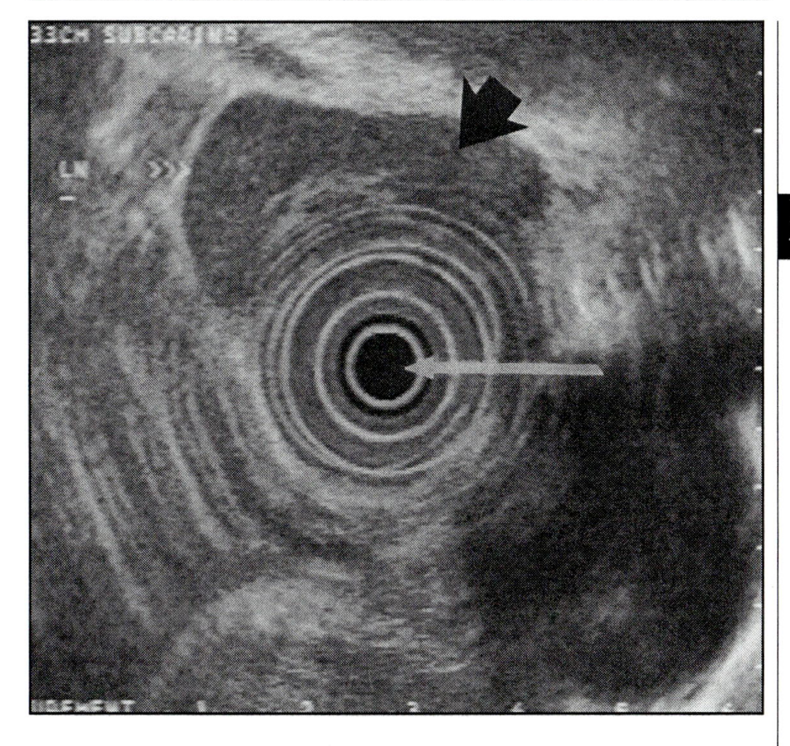

Figure 22.15. A large subcarinal lymph node (black arrow) in the mediastinum in a patient with lung cancer, as seen with a radial endoscopic ultrasound instrument. White Arrow = radial EUS transducer in the esophagus.

- N1:— Metastases in 1 to 3 regional lymph nodes
- N2:— Metastases in >4 regional lymph nodes

Lung Carcinoma

- Endoscopic ultrasound is unable to provide primary tumor (T) staging for lung carcinoma
- EUS can help in assessing lymph nodes in the mediastinum for metastases with lung carcinoma (N stage)
- N-staging in lung carcinoma according to the TNM classification is as follows:
  - NO: No regional lymph node metastasis
  - N1: Metastases in ipsilateral peribronchial and/or ipsilateral hilar lymph nodes and/or intrapulmonary nodes including involvement by direct extension
  - N2: Metastases in ipsilateral mediastinal and/or subcarinal lymph nodes
  - N3: Metastases in contralateral mediastinal, contralateral hilar, ipsilateral or contralateral scalene or supraclavicular lymph nodes.
- EUS cannot assess lymph node invasion at all of the above locations in lung cancer.

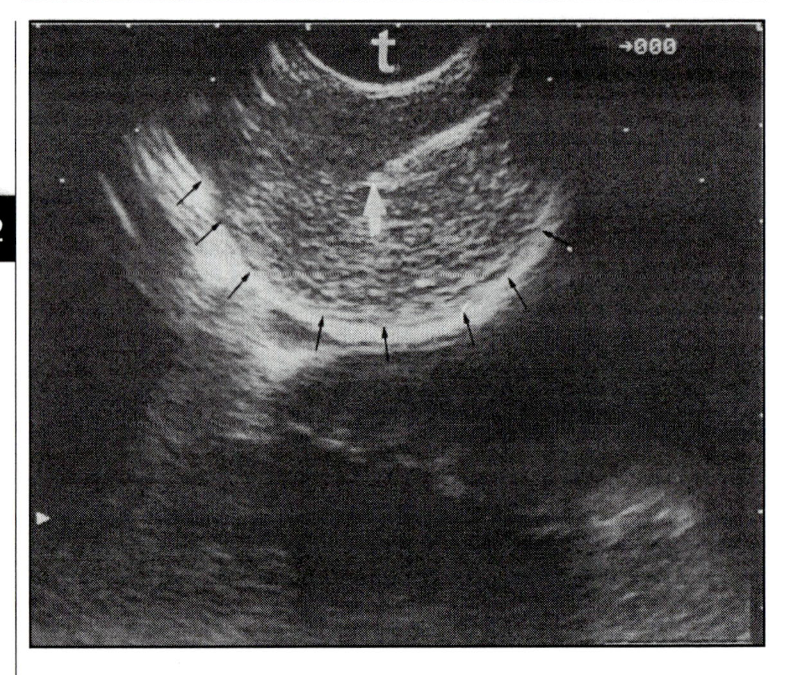

Figure 22.16. Linear EUS guided transesophageal fine needle aspiration of a large mediastinal lymph node in a patient with nonsmall cell lung cancer. White Arrow = needle within the lymph node, Black Arrows = Lymph Node, T = Linear Array Transducer in the Esophagus.

- Mediastinal lymph nodes that are located around the esophagus are accessible by EUS e.g., subcarina, aortapulmonary window, etc.
  - Imaging of these mediastinal lymph nodes alone is not a reliable indicator for presence of malignant invasion. (Fig. 22.15)
  - Definite diagnosis of malignant invasion in mediastinal lymph nodes can be made with EUS guided real-time transesophageal fine needle aspiration of these lymph nodes in patients with lung cancer.[4,6,7] (Fig. 22.16)

Accuracy of Endoscopic Ultrasound for Tumor Staging

- For esophageal, gastric and rectal carcinomas the accuracy of EUS for primary tumor (T) staging is about 90%.[8,9,10]
- The accuracy of lymph node (N) staging by EUS in esophageal, gastric and rectal cancer by imaging alone is 60-80%.[8-10] However, with the addition of EUS guided fine needle aspiration, the accuracy of EUS is significantly enhanced for diagnosis of malignant invasion of a lymph node.[4]
- In lung cancer lymph node staging, if the mediastinal lymph nodes are enlarged and in a location accessible by tranesophageal EUS (subcarina, aortopulmonary window), the accuracy of EUSFNA for detecting malignant lymph node invasion is high and in the range of 84-100%.[4,6,7]

Complications

- EUS overall is a safe test. In experienced hands, EUS by imaging alone does not appear to have an increased risk of complications over a standard upper endoscopy.
- In a large multicenteric study,[11] the overall complication rate of EUS was 1.7% with a severe complication rate of 0.3% in patients undergoing EUS. Complications in this study included:
  - mild complications: Sore throat, oxygen desaturation during EUS, intravenous site phlebitis
  - moderate complications: abdominal pain requiring short hospitalization, severe, esophageal perforation
- Even when fine needle aspiration is added to EUS, the overall complication rate is low and around 1.1%.[12]
- In conclusion, EUS plays a significant role in staging esophageal, gastric, rectal and lung cancer.

Acknowledgement

Many thanks to Vicky Rhine for her assistance in typing this manuscript.

Selected References

1. Kimmey MB, Martin RW, Haggit RC et al. Histologic Correlates of Gastrointestinal Ultrasound Images. Gastroenterology 1989; 96:433-441.
2. Hawes RH. Normal Endosonographic Findings. Gastrointest Endosc 1996; 43:S6-S10.
3. Catalano MF, Sivak MV Jr, Rice T et al. Endosonographic features predictive of lymph node metastases. Gastrointest Endosc 1994; 40:442-446.
4. Bhutani MS, Hawes RH, Hoffman BJ. A comparison of the accuracy of echo features of lymph nodes during endoscopic ultrasound (EUS) and EUS-guided fine needle aspiration for diagnosis of malignant lymph node invasion. Gastrointest Endosc 1997; 45:474-479.
5. Hermanek P, Hutter RVP, Sobin LH et al. TNM Atlas. 4th Ed. Berlin/Heidelberg: Springer, 1997.
6. Gress F, Savides TJ, Sandler A et al. Endoscopic ultrasonography, fine–needle aspiration guided by endoscopic ultrasonography and computed tomograhpy in the preoperative staging of non-small cell lung cancer. A comparison study. Ann Intern Med 1997; 127:604-616.
7. Silvestri GA, Hoffman BJ, Bhutani MS et al. Endoscopic ultrasound with fine needle aspiration in the diagnosis and staging of lung cancer. Ann Thoracic Surg 1996; 61:1441-1446.
8. Caletti G, Odegaard S, Rosch T et al. Endoscopic ultrasonography (EUS) : A summary of the conclusions of the working party for the Tenth World Congress of Gastroenterology, Los Angeles, California, October 1994. Am J Gastroenterol 1994; 89:5138-5143.
9. Lightdale CJ. Staging of Esophageal Cancer 1: Endoscopic ultrasonography. Sem Oncology 1994; 21:438-446.
10. Boyce GA, Sivak MV Jr, Lavery IC et al. Endoscopic ultrasonography in the preoperative staging of rectal carcinoma. Gastrointest Endosc 1992; 38:468-471.
11. Nickl NJ, Bhutani MS, Catalano M et al. Clinical implications of endoscopic ultrasound: The American Endosonography Club Study. Gastrointest Endosc 1996; 44:371-377.
12. Wiersema MJ, Vilmann P, Giovannini M et al. Endosonography-guided fine-needle aspiration biopsy: Diagnostic accuracy and complication assessment. Gastroenterology 1997; 112:1087-1095.

CHAPTER 23

Endoscopic Ultrasound: Submucosal Tumors and Thickened Gastric Folds

Kenji Kobayashi and Amitabh Chak

Submucosal Tumors

- Intraluminal bulge covered with normal appearing mucosa
- Extrinsic lesions which can produce the same endoscopic appearance
  - EUS evaluation of submucosal tumors
    - Extraluminal compression can be easily differentiated from submucosal tumor by EUS because the normal five-layer gastrointestinal wall structure is preserved and adjacent organ (e.g., spleen, liver, pancreas) causing the bulge can be identified
    - When intraluminal lesions are evaluated by EUS, it is important to define the sonographic layer of origin, echo pattern, echo texture, internal features, and margin of the tumor as these features are helpful in differential diagnosis of submucosal tumors
  - Characteristic EUS findings of common submucosal tumors
    - **Benign stromal tumors**
      - Benign stromal tumors are the most commonly encountered
      - Typically appear as homogeneous hypoechoic lesion with smooth margin
      - Typically arise in the muscularis propria (layer 4), although small leiomyomas may arise from the muscularis mucosae (layer 2)
      - Rarely, an inhomogenous echo pattern or irregular margin can be observed the term leiomyoma is commonly used to describe these lesions, but is not appropriate because these tumors are not of smooth muscle origin
    - **Malignant stromal tumors**
      - Malignant stromal tumors are typically large (>4 cm), inhomogenous, hypoechoic lesion with irregular margins
      - Also arise from the forth layer
      - Often associated with central hypoechoic/anechoic areas, postulated to represent liquefaction necrosis
      - Central hypoechoic areas may be seen in large (>5 cm in diameter) benign stromal tumors, but are more common in malignant ones
      - Invasion of adjacent organs is also suggestive of malignancy
      - Although small, homogenous, hypoechoic, regular tumors are almost always benign, the treatment of these tumors based on EUS imaging is still not defined. Biopsies may not help distinguish benign and malignant tumors. Even after surgical resection, pathologic classification of benign and malignant tumors can be difficult.

Gastrointestinal Endoscopy, edited by Jacques Van Dam and Richard C. K. Wong.

- **Lipomas**
  - Characteristically hypoechoic and homogeneous with smooth margins
  - Typically arise in the third sonographic layers (submucosa)
  - Usually benign, most lipomas are left alone unless they are large enough to cause bleeding or obstruction
- **Pancreatic rest**
  - Variable EUS features
  - May be homogeneous, hypoechoic lesions within the third layer, or they may be hypoechoic lesion with fine, scattered hyperechoic foci
  - Occasionally, duct-like structures may be seen within these lesions
  - Generally originate from the submucosa (layer 3), but occasionally extend to the mucosa (layers 1 and 2) or muscularis propria (layer 4). Pancreatic rests are generally found incidentally and do not require any intervention.
- **Cysts**
  - Readily identified as well-demarcated, rounded, anechoic lesions confined within the third layer (submucosa)
  - Small cysts can be compressed with the ultrasound endoscope balloon, and thus care needs to be taken not to overinflate the balloon when imaging these lesions
  - Ultrasound catheter probes, which can be passed through the accessory channel of endoscopes, are ideally suited for imaging small cysts
  - Most occur in the esophagus
  - Most commonly from congenital duplication but may also be related to ectopic gastric mucosa or ectopic bronchial mucosa
  - Large symptomatic cysts should be treated with surgical excision or marsupialization. Endoscopic or EUS guided needle aspiration can be performed in patients who acutely symptomatic, especially if trachea is being compressed.
- **Extrinsic compression**
  - Aorta, liver, gallbladder, spleen, and splenic artery or splenic vein may compress the gastrointestinal tract when enlarged or even when of normal size. In these cases, the normal five-layer pattern of the gastrointestinal wall is preserved and the organ causing compression is easily identified.
  - ***Esophageal compression*** can be secondary to reactive adenopathy, mediastinal granulomas, metastatic adenopathy, mediastinal tumors, lung tumors, or enlarged left atrium
  - ***Gastric compression*** may commonly be caused by pancreatic pseudocysts, pancreatic tumors, retroperitoneal soft tissue tumors, primary or metastatic hepatic tumors, or splenomagaly
  - ***Duodenal compression*** can be seen in association with pseudocysts or tumors in the pancreatic head
  - ***Rectal compression*** may commonly be related to tumors of the prostate or from uterine tumors. Extrinsic inflammatory as well as malignant processes that cause extrinsic compression may involve the fifth and forth sonographic layers.

Table 23.1. Differential diagnosis of thickened gastric folds

- Hyper-rugosity
- Gastritis
- Ménetriér's disease
- *Helicobacter pylori* gastritis
- Zollinger-Ellison syndrome
- Gastric varices
- Eosinophilic gastritis
- Granulomatous gastritis
    - Infectious
        - Tuberculosis
        - Histoplasmosis
        - Secondary syphilis
        - Parasitic infection (e.g., Anisakiasis)
        - Whipple's disease
    - Noninfectious
        - Crohn's disease
        - Sarcoidosis
        - Allergic granulomatosis
    - Malignancies
        - Gastric cancer
        - Lymphoma
        - MALToma
        - Linitis plastica due to metastatic cancer (especially breast cancer)

Thickened Gastric Folds

- Folds greater than 10 mm on barium contrast x-ray or as those that do not flatten with air insufflation at endoscopy
- May appear as a thickened on CT scan, although often an artifact of insufficient gastric distention

Etiology

- Ranges from a normal variant (hyper-rugosity) to benign inflammatory conditions such as gastritis to malignancy, Table 23.1

Normal EUS of the Stomach

- Upper limit of thickness for the gastric wall is approximately 4 mm in the antrum and 3 mm in the body and fundus
- Degree of luminal distention with water and balloon compression when performing EUS can affect layer visualization and wall thickness measurement. The size ratio of layers 2, 3, and 4 is roughly 1:1:1.

EUS of Thickened Gastric Folds

- Majority of patients with thickening limited to layer 2 (deep mucosa) have a benign condition
- Thickening involving layer 4 is highly suggestive of malignant process

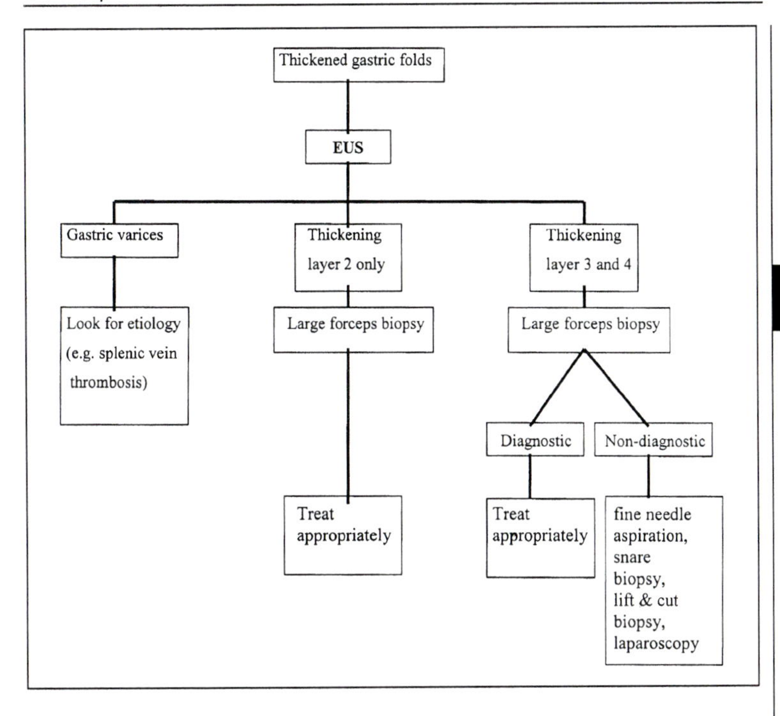

Figure 23.1. Diagnostic evaluation of thickened gastric folds.

- Although EUS can characterize the features of thickened gastric folds, it should be remembered that EUS can not often make a definitive diagnosis. Figure 23.1 shows a diagnostic approach to thickened gastric folds
  - **Gastric varices**. Easily identified during EUS as anechoic serpiginous structures. EUS is recommended prior to large cup forceps biopsy or snare biopsy as a biopsy of a varix would obviously result in somewhat adverse consequences.
  - **Ménetriér's disease**. Rare disorder characterized by large gastric folds, foveolar hyperplasia, atrophy of glands, and hypochlorhydria. On EUS, the second layer (deep mucosa) is symmetrically and homogenously enlarged with small cystic spaces within it. Patients may be asymptomatic or may have anemia and edema related to protein losses. Since the etiology is unknown, the treatment is generally to treat symptoms.
  - **Gastritis**. Acute or chronic infection with Helicobacter pylori may cause thickened gastric folds. EUS demonstrates thickening of layer 13 (mucosa and submucosa) in patients with chronic active gastritis due to *H. pylori*. The wall thickening can be resolved upon eradication of *H. pylori* and resolution of gastritis.
  - **Linitis plastica**. Primary or metastatic carcinoma to the stomach may produce submucosal infiltration and fibrosis resulting in linitis plastica. At EUS, a diffusely hypoechoic or inhomogenous wall thickening characterize linitis

plastica. Although layer 2 can be involved, especially deeper layer involvement (layers 3 and 4) is especially characteristic.
- Stomach may be nondistensible when water is infused.
- Differentiation from lymphoma is sometimes problematic. One study reported that the mucosa was significantly thicker in lymphoma than in linitis plastica, while the submucosa was thicker in carcinoma than lymphoma.
- Fine needle aspiration (FNA) of the thickened gastric folds or adjacent lymph nodes can be helpful for diagnosis when large particle biopsy fails
- **Lymphoma/MALToma**
  - Tumors of mucosal associated lymphoid tissue (MALTomas) and primary gastric lymphomas can present with various endoscopic and endosonographic appearances that are quite similar requiring histology for definitive diagnosis. MALToma may be confined to the mucosal surface (first two layers) or may extend to all layers
  - Endosonographic staging may be useful in predicting response to antibiotic therapy
  - Cases with deeper involvement are generally unresponsive and may require more aggressive chemotherapy
  - Primary gastric lymphomas are typically characterized by a diffuse or localized hypoechoic thickening of layer 2-4. EUS can assess depth of invasion as well as extent of extragastric involvement.

Selected References

1. Tio TL, Tytgat GNJ, den Hartog Jager FCA. Endoscopic ultrasonography for the evaluation of smooth muscle tumors in the upper gastrointestinal tract: An experience with 42 cases. Gastrointest Ensosc 1990; 36:342-350.
2. Yasuda K, Nakajima M, Yoshida S et al. The diagnosis of submucosal tumors of the stomach by endoscopic ultrasonography. Gastrointest Endosc 1989; 35:10-15.
3. Boyce GA, Sivak MV, R+sch T et al. Evaluation of submucosal upper gastrointestinal tract lesions by endoscopic ultrasound. Gastrointest Endosc 1991; 37:449-454.
4. Rösch T, Lorenz R, Dancygier H et al. Endosonographic diagnosis of submucosal upper gastrointestinal tract tumors. Scand J Gastoenterol 1992; 27:1-8.
5. Mendis RE, Gerdes H, Lightdale CJ et al. Large gastric folds: A diagnostic approach using endoscopic ultrasonography. Gastrointest Endosc 1994; 40:437-441.
6. Avunduk C, Navah F, Hampf F et al. Prevalence of Helicobacter pylori infection in patients with large gastric folds: Evaluation and follow-up with endoscopic ultrasound before and after antimicrobial therapy. Am J Gastroenterol 1995; 90:1969-1973.
7. Songnr Y, Okai T, Watanabe H et al. Endoscopic evaluation of giant gastric folds. Gastrointest Endosc 1995; 41:468-474.

CHAPTER 24

Endoscopic Ultrasonography (EUS) of the Upper Abdomen

Shawn Mallery

Background

- Basic principles of ultrasound imaging. Ultrasound imaging is based upon the same principles as SONAR, RADAR and echolocation (used by bats and dolphins). The speed of sound is relatively constant in any given media. By measuring the time required for sound to reflect off a distant object and return to the observer, the distance to that object can be determined (Distance = Velocity x Time). For the purposes of imaging, each echo is displayed visually as a bright spot at the measured distance from the source of sound (the ultrasound transducer).
- Ultrasound characteristics of various tissues. Structures containing fluid produce few, if any echoes and are thus displayed as dark black (anechoic). Solid structures with a high water content appear dark but contain scattered internal echoes (hypoechoic). Solid structures with a high fat content appear relatively white (hyperechoic).
- Shadowing. When structures are dense or highly reflective and do not transmit sound, only the surface facing the transducer can be visualized. As little sound energy is transmitted through to deeper structures, these are hidden.
- Acoustic enhancement. Entirely fluid-filled structures may transmit sound more readily than surrounding solid tissue. Structures deeper than the fluid are therefore exposed to more energy and appear brighter (reflect more energy) than adjacent structures that are deep to solid tissue.
- Limitations of traditional ultrasonography
  - Tissue-related limitations. Sound waves are transmitted more efficiently through fluid than solids or gases. Ultrasound imaging is thus not possible through dense solids (e.g., bone) or gases (e.g., lungs, bowels). For this reason, standard ultrasound imaging of the mediastinum, distal common bile duct (which passes posterior to the duodenum) and pancreas has been of limited value.
  - Distance-related limitations. As sound frequency (measured in megahertz–MHz) increases, improved imaging resolution is obtained. Unfortunately, higher frequency sound is transmitted less readily by tissue, limiting depth of imaging.

Rationale for Endoscopic Ultrasound

- Avoidance of bone and gas-filled structures. Placement of the ultrasound transducer within the gastrointestinal lumen and, when necessary, surrounding the

Gastrointestinal Endoscopy, edited by Jacques Van Dam and Richard C. K. Wong.

transducer with water allows interfering gas and bone to be avoided. Fluid may be used for acoustic coupling either by inflating a water-filled balloon around the echoendoscope or instilling water into the gastrointestinal lumen.

- Use of high-resolution (high-frequency) imaging. Placement of the transducer in close proximity to the lesion of interest allows the use of higher frequency sound (see above). This provides extremely detailed imaging not possible with transabdominal ultrasound.

Endoscopic Ultrasound Equipment

- The building block – piezoelectric crystals. Sound energy is produced via unique crystals that release sound when exposed to electrical current. These same crystals produce an electrical current when struck by returning echoes. This electrical current is translated into an image by a computer.
- Radial echoendoscopes. Radial echoendoscopes contain ultrasound transducers that rotate in a circular plane perpendicular to the long axis of the endoscope. This produces an image perpendicular to the long axis of the echoendoscope, with the scope centered in the imaging field. With the scope parallel to the spine, images obtained with radial devices resemble CT scan.
- Linear array echoendoscopes. Linear echoendoscopes contain a fixed transducer which produces a sector-shaped image aligned parallel to the long axis of the endoscope. The endoscope is displayed at the top of the image, similar to images obtained by transabdominal ultrasound. Most of these devices also allow Doppler imaging, in which moving material (e.g., blood) is displayed in color.
- Mechanical array echoendoscope. Mechanical array echoendoscopes produce images via the use of a rotating acoustic mirror. The resulting image is aligned along the long axis of the scope and wraps around the scope tip. Doppler imaging is not possible with currently available equipment.
- Specialized instruments
  - Miniprobes. Miniature ultrasound probes are available which can be placed through the instrument channel of a standard, non-ultrasound endoscope. These probes, because of their small diameter, can be advanced inside the bile duct and pancreatic duct.
  - Esophagoprobe (Olympus MH908). This device is narrower than standard echoendoscopes and has a smoothly tapered tip. It is designed to allow imaging in the setting of esophageal obstruction. A guidewire is advanced through the obstruction (either endoscopically or via fluoroscopic-guidance). The esophagoprobe is then gently advanced over the wire and through the narrowing. This is important for the staging of esophageal carcinoma, for which the presence of metastases in celiac lymph nodes is of prognostic importance. No optical imaging is provided.

Upper Abdominal Anatomy

- Gastrointestinal wall layers (Fig. 24.1). One of the primary advantages of EUS compared to other imaging modalities is its unique ability to visualize the individual histologic wall layers of the gastrointestinal tract. This is important for evaluating depth of tumor infiltration, diseases which cause wall thickening of specific layers (e.g., lymphoma, Ménétrier's disease) and intramural tumors. A five-layered wall structure is visualized throughout the GI tract as alternating concentric hyperechoic and hypoechoic regions. *(NOTE: With radial imaging, several bright rings are also artifactually produced by the echoendoscope itself. These are closer to the endoscope, perfectly circular and concentric with the endoscope axis. These should not be confused as representing wall layers.)*

- Layer 1—mucosa (hyperechoic). The first bright band surrounding the echoendoscope represents the echoes produced by the luminal surface and superficial mucosa.
- Layer 2 – deep mucosa (hypoechoic). The next layer encountered is a dark band corresponding to the deep layers of the mucosa (primarily the muscularis mucosa). High frequency (e.g. 20 MHz) imaging can often separate this into additional layers representing the mucosa, muscularis mucosa and the interface between these two layers.
- Layer 3 – submucosa (hyperechoic). The submucosa is seen as a bright white structure due to its high fat content.
- Layer 4 – muscularis propria (hypoechoic). This layer is seen histologically as an inner circular layer of muscle and an outer longitudinal layer, separated by a thin layer of connective tissue. These three individual layers can occasionally be visualized, however typically these are seen as a single hypoechoic band.
- Layer 5 – adventitia/serosa (hyperechoic). The esophagus and retroperitoneal portions of the GI tract are surrounded by fat and connective tissue (the adventitia). Intraperitoneal structures are surrounded by a thin, hyperechoic layer of connective tissue (the serosa.)

- Pancreas (Fig. 24.2). The pancreas is visualized as a structure of mixed echogenicity ("salt and pepper"). The body and tail are seen through the posterior wall of the stomach. The head and uncinate are generally best viewed through the medial wall of the duodenum. The pancreatic duct can usually be seen within the pancreatic parenchyma as a hypoechoic tubular structure.
- Extrahepatic biliary tree (Fig. 24.3). The extrahepatic biliary tree can typically be visualized in its entirety via EUS. The common bile duct passes superiorly from the major papilla through the pancreatic head and posterior to the duodenal bulb. At this point it is roughly parallel to the portal vein, which lies further from the duodenal lumen. Superior to the bulb, the cystic duct branches and leads to the gallbladder. Above the cystic duct, the bile duct is termed the common hepatic duct. This then branches at the hepatic hilum and enters the liver.
- Arterial structures of the upper abdomen
  - Aorta. The aorta is readily visualized posterior to the stomach. With radial imaging, it appears as a circular, anechoic/hypoechoic structure that remains relatively constant in size and location with scope insertion. With linear imaging, it is seen as a horizontal band across the mid-portion of the image.
  - Celiac trunk. The celiac trunk is the first major branch to arise from the intra-abdominal aorta. This is typically encountered approximately 45 cm from the teeth. After a short segment, the trunk splits into the three major vessels – the splenic artery (which travels to the left into the splenic hilum), the hepatic artery (which can be followed into the hepatic hilum) and the left gastric artery (which follows the lesser curvature of the stomach). *(NOTE: With radial imaging, the diaphragmatic crus can sometimes mimic the celiac trunk. These structures can be distinguished by the fact that the crus drapes over the aorta but never joins it.)*

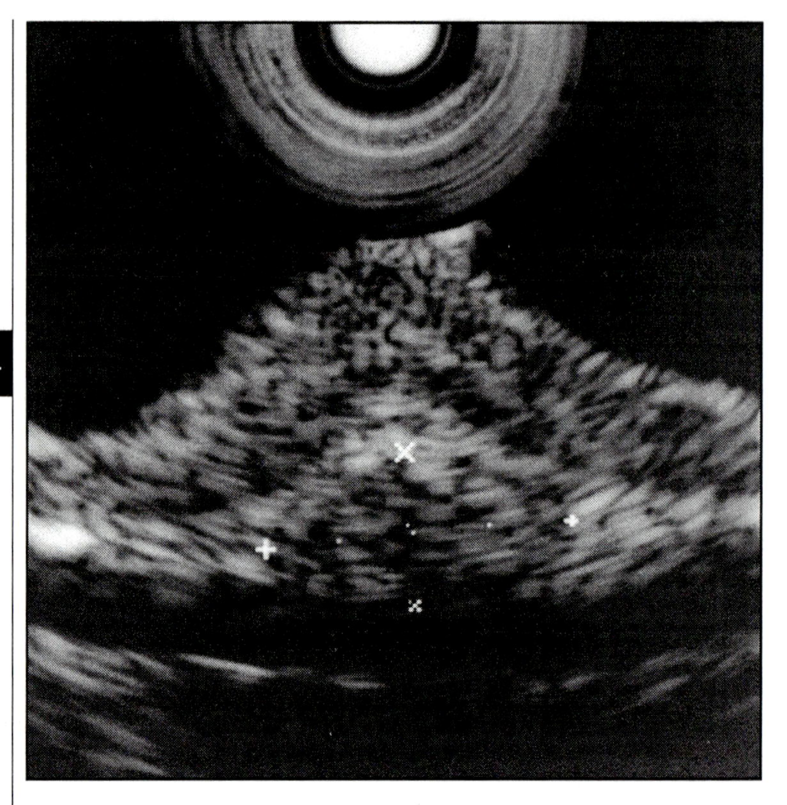

Figure 24.1. High frequency (20 MHz) EUS of gastric carcinoid. A well circumscribed, ovoid hypoechoic lesion is present within the submucosa that deforms the overlying mucosa. The lesion measures only 2 x 4 mm. There is no abnormality in the underlying muscularis propria to suggest invasion. This is a good example of walllayer visualization by EUS. In this high resolution image, the mucosa has a suggestion of two distinct hypoechoic bands, corresponding to the mucosa and muscularis mucosa separated by a hyperechoic border echo resulting from the interface of these two structures. (Imaged via Olympus UM3R 20 MHz probe, Olympus America Inc., Melville, NY)

- Superior mesenteric artery (SMA). The SMA is the second major vessel to arise from the aorta. It is seen 12 cm distal to the celiac trunk. No significant branches are typically seen to arise from the SMA by EUS.

- Venous structures of the upper abdomen
  - Portal system (Fig. 24.2). The splenic vein lies along the posterior surface of the pancreatic body and tail. To the right of midline, it joins the superior mesenteric vein (at the "portal confluence") to form the portal vein. The portal vein then passes superiorly into the hepatic hilum.

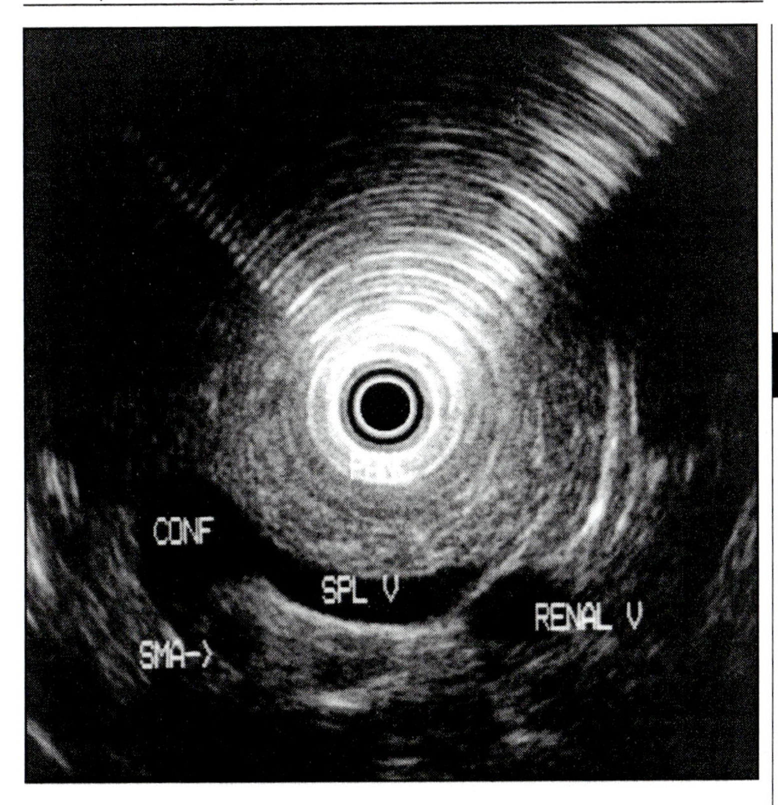

Figure 24.2. Normal radial EUS of the pancreatic body. The echoendoscope (black circle in middle of the image) is positioned against the posterior gastric wall. The pancreas is seen as tissue of mixed echogenicity ("salt and pepper") bordered posteriorly by the anechoic (black), club-shaped splenic vein (SPL V) and portal confluence (CONF). The superior mesenteric artery (SMA) is a round vessel with hyperechoic walls which indents the portal confluence. The pancreatic duct is not seen in this image. The radiating concentric white arcs in the upper half of the image are due to air artifact. (Imaged at 7.5 MHz via Olympus GFUM130 videoechoendoscope, Olympus America Inc, Melville, NY)

- Inferior vena cava (IVC). The IVC is a long, straight, large caliber vessel that passes deeply posterior to the pancreatic head to enter the posteroinferior portion of the liver. The three hepatic veins join the IVC before it exits the liver to empty into the right atrium.

- Lymph nodes. Lymph nodes are visualized as round, isoechoic to hypoechoic structures. They are distinguished from vessels by observing the effects of small movements of the scope over the structure. Nodes will disappear and reappear ("wink"), whereas vessels will persist, elongate or branch.

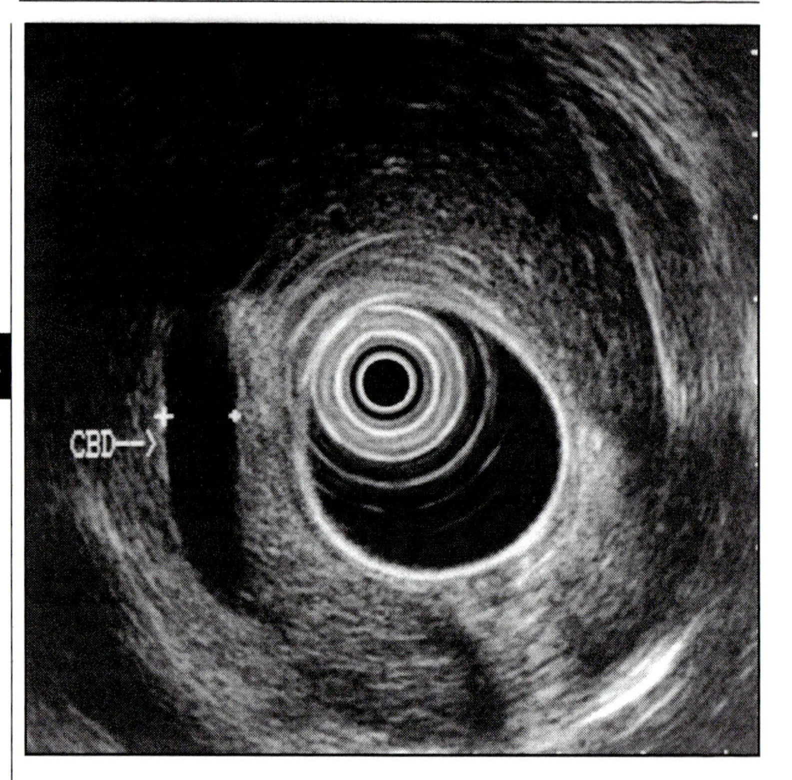

Figure 24.3. Normal radial EUS of the common bile duct (CBD). The echoendoscope is positioned in the duodenal bulb. The CBD, here measuring 8.5 mm in diameter, passes posterior to the bulb to enter the pancreas. The liver is seen above the transducer in this view as a hypoechoic triangular region containing numerous small anechoic ducts. With subtle movements of the transducer, the CBD can be traced to the major papilla. (Imaged at 7.5 MHz via Olympus GFUM130 videoechoendoscope, Olympus America Inc, Melville, NY)

- Common locations for lymph nodes. Nodes are frequently seen adjacent to the celiac trunk (celiac nodes), the gastric wall (perigastric nodes), the pancreas (peripancreatic nodes), the hilum of the liver (perihilar nodes), and just distal to the diaphragm between the liver and lesser curve of the stomach (gastrohepatic ligament nodes).
- Endosonographic characteristics of malignant nodes. Malignant nodes tend to be large (> 1 cm in diameter), round, relatively hypoechoic (compared to surrounding tissue) and sharply demarcated. Benign nodes tend to be smaller, flat, oblong or triangular, isoechoic and have hazy, indiscrete borders. The likelihood of malignancy approaches 100% when all four malignant characteristics are present.[1]

- Liver (Fig. 24.3). The liver is a large, hypoechoic structure seen best along the lesser curvature of the stomach. It is readily identified by the numerous branching vessels and ducts within the parenchyma. Typically, only the medial 2/3 of the liver can be visualized via EUS, although this varies depending upon patient size and scanning frequency. The echogenicity of the liver and spleen should be nearly identical. A liver that is more hyperechoic (bright) suggests fatty infiltration.
- Spleen. The spleen is another large hypoechoic structure adjacent to the gastric wall. It can occasionally be confused for the liver, however it is distinguished by the lack of internal branching structures.

Endosonography of Pancreatic Malignancies

- EUS is highly sensitive for **small pancreatic masses**. Patients with pancreatic mass lesions frequently present with evidence of biliary obstruction. Often, no mass is seen by traditional radiologic studies. EUS is highly sensitive for small masses,[2] and at the same time allows tissue sampling and staging to be performed.
- EUS appearance of pancreatic carcinoma. **Pancreatic adenocarcinoma** is seen a focal hypoechoic region within the pancreas.
- **EUS-guided fine-needle aspiration** (FNA) (Fig. 24.4). Not all pancreatic masses arise due to adenocarcinoma. The differential diagnosis includes focal chronic pancreatitis, neuroendocrine tumors, lymphoma, metastases and several rarer tumors of pancreatic origin. Using linear or mechanicalarray imaging, a thin (1922 gauge) needle can be advanced into the mass under sonographic guidance at the time of initial EUS. Tissue is then aspirated for cytologic analysis. Several centers have reported a diagnostic accuracy of 80-90% for this technique with rare complications.[3]
- EUS staging of pancreatic adenocarcinoma
  - T-stage (Figs. 24.4, 24.5)
    - T-stage is determined by the extent of invasion of the primary tumor into surrounding structures. Clinically, the most important structure is the portal vein, as involvement generally precludes curative surgical resection. Options for T-staging include EUS, CT, MRI and angiography. Several studies have shown EUS to be superior to these other modalities for the detection of portal vein involvement.[4]
    - Tx Primary tumor cannot be assessed
    - T0 No evidence of primary tumor
    - Tumor limited to the pancreas
      - T1a Tumor ≤ 2 cm in greatest dimension
      - T1b Tumor > 2 cm in greatest dimension
    - T2 Tumor extends directly to duodenum, bile duct or peripancreatic tissues
    - T3 Tumor extends directly into stomach, spleen, colon or adjacent "large" blood vessels *(NOTE: These can be remembered as organs not removed via Whipple resection)*
  - N-stage. See section F.3 for a discussion of EUS characteristics of malignant nodes.
    - Nx Regional lymph nodes cannot be assessed
    - N0 No regional lymph node metastasis

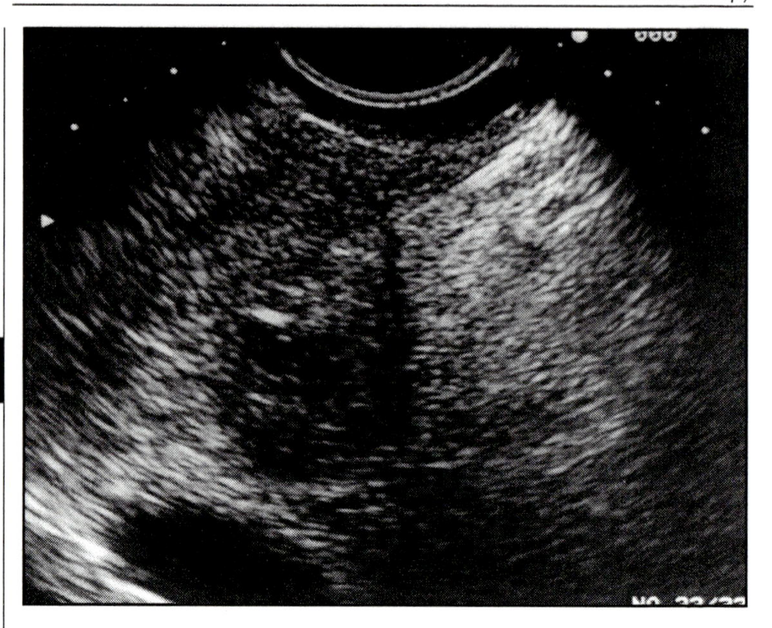

Figure 24.4. EUS-guided fine needle aspiration. The echoendoscope is positioned to image through the posteroinferior surface of the duodenal bulb. A hypoechoic mass measuring 17 x 22 mm is seen in the center of the picture. A 22 gauge needle can be seen as a bright line entering the mass from the upper right of the image, with the tip positioned in the center of the mass. Cytology confirmed adenocarcinoma. This image shows sparing of the portal vein (seen at the lower left) but involvement of the duodenal wall (the tumor extends up to abut the surface of the water-filled balloon) – stage T2. The patient refused surgery and presented 1 month later with duodenal obstruction. (Imaged at 5 MHz via Pentax FG32UA linear array echoendoscope, Pentax Corp, Orangeburg, NY).

- N1 Regional lymph node metastasis
  - N1a Metastasis in a single regional lymph node
  - N1b Metastasis in multiple regional lymph nodes
- M-stage. Common sites of metastasis include the liver and peritoneum (with ascites). EUS can detect and biopsy extremely small intrahepatic lesions or fluid collections, but typically cannot visualize the entire liver.
  - Mx Distant metastasis cannot be assessed
  - M0 No distant metastasis
  - M1 Distant metastasis

• Endosonographic localization of pancreatic endocrine tumors. **Endocrine tumors of the pancreas** (insulinoma, gastrinoma, glucagonoma and others) may cause debilitating hormonal symptoms while still extremely small. Surgical resection is often curative, but localization may be quite difficult even at the time of laparotomy. EUS has been reported to be the most accurate test for the preoperative localization of pancreatic endocrine tumors.[5]

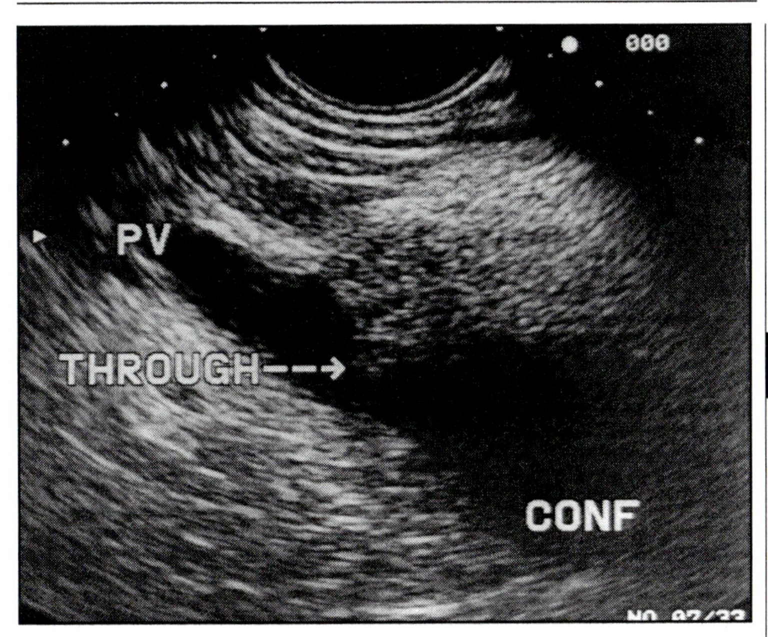

Figure 24.5. Portal vein invasion. The echoendoscope is positioned to image through the posteroinferior surface of the duodenal bulb. A hypoechoic mass is seen protruding into the lumen of the portal vein (PV), which passes from the upper left to lower right corner of the image (CONF = portal confluence). (Imaged at 5 MHz via Pentax FG32UA linear array echoendoscope, Pentax Corp, Orangeburg, NY).

Endosonography of Cystic Pancreatic Lesions

- Differential diagnosis of intrapancreatic fluid collections. Pancreatic fluid collections most commonly are pseudocysts. In patients without a history of severe acute pancreatitis, however, several cystic neoplasms should be considered. These include serous cystadenomas (benign), mucinous cystadenomas (premalignant), mucinous cystadenocarcinomas (malignant) and several other lesions. Unlike the liver and kidney, simple congenital cysts are rare in the pancreas.
- Endosonographic appearance of **pancreatic cysts**. Fluid collections appear as round or oval anechoic structures with a bright back surface. Some cysts contain internal debris or fibrous septa. An outer wall, when visualized, may be of variable thickness. The presence of an adjacent mass is suspicious for neoplasm, however cyst morphology cannot reliably diagnose or exclude malignancy.
- EUS-guided cyst aspiration. When entities other than pseudocyst are suspected, fluid can be aspirated by inserting a needle under EUS guidance. Fluid is generally submitted for cytology, however the accuracy of cytology alone is relatively low. Several tumor markers are being investigated to improve this yield. Antibiotic prophylaxis (e.g., ciprofloxacin 500 mg po bid x 5 days) is generally recommended to prevent infection after aspiration.

- EUS-guided pseudocyst drainage. In patients with symptomatic pseudocysts that fail to resolve after 68 weeks, endoscopic or surgical drainage may be recommended. EUS can be useful both in excluding the presence of blood vessel within the proposed drainage track and in localizing the pseudocyst when a bulge is not seen endoscopically. Prototype instruments are being developed to allow direct placement of drains through the EUS scope.

Endosonographic Diagnosis of Chronic Pancreatitis

- Histopathologic changes of chronic pancreatitis. Chronic pancreatitis is characterized by infiltration of the gland by chronic inflammatory cells and fibrous tissue. In moderate to severe disease, there may be deposition of calcium, pseudocyst formation, pancreatic ductal stricture and/or the formation of intraductal stones.
- Parenchymal changes (Fig. 24.6). The following changes are associated with chronic pancreatitis: hyperechoic foci or shadowing microcalcifications, hypoechoic foci, hyperechoic fibrotic strands and/or an irregular outer pancreatic border.
- Ductal changes (Fig. 24.7). Changes in the main pancreatic duct associated with chronic pancreatitis include: ductal dilation, intraductal stones, hyperechoic duct wall and visible side-branches
  - The greater the number of ductal and parenchymal changes, the more likely the diagnosis of chronic pancreatitis
  - In general, three or more findings are required for this diagnosis.[6]
  - EUS appears to be more sensitive than ERCP for mild disease and equal to ERCP for moderate to severe disease
  - EUS can identify parenchymal changes of chronic pancreatitis in patients with normal-appearing pancreatic ducts by ERCP. This has been suggested as evidence that EUS is more sensitive for early disease; however it is possible that EUS classifies some patients as abnormal who do not have disease. This will require further study.

Endosongraphic Diagnosis of Choledocholithiasis

- EUS appearance of common duct stones. The common bile duct can be visualized in its entirety by EUS (Fig. 24.3). Intraductal stones are seen as hyperechoic densities within the duct which shadow (similar to the pancreatic stone seen in Fig. 24.7)
- Sensitivity and accuracy of EUS for common duct stones. The reported sensitivity and accuracy rates of EUS are roughly equivalent to that of ERCP and MRCP (90-97%).[7]
- Advantages of EUS. The major advantages of EUS are a reduced risk of procedure-related pancreatitis and the avoidance of ionizing radiation needed for ERCP
- Disadvantages of EUS. The main disadvantage of EUS is the inability to remove stones when found.
- Indications for EUS in reference to ERCP. EUS should be considered when there is a low to intermediate suspicion of common duct stones, a high-risk of ERCP-related pancreatitis (e.g., those recovering from a recent bout of acute pancreatitis or a previous history of ERCP-induced pancreatitis) or a relative contraindication to the use of ionizing radiation (e.g., pregnancy). Early studies suggest that this is a cost-saving strategy.

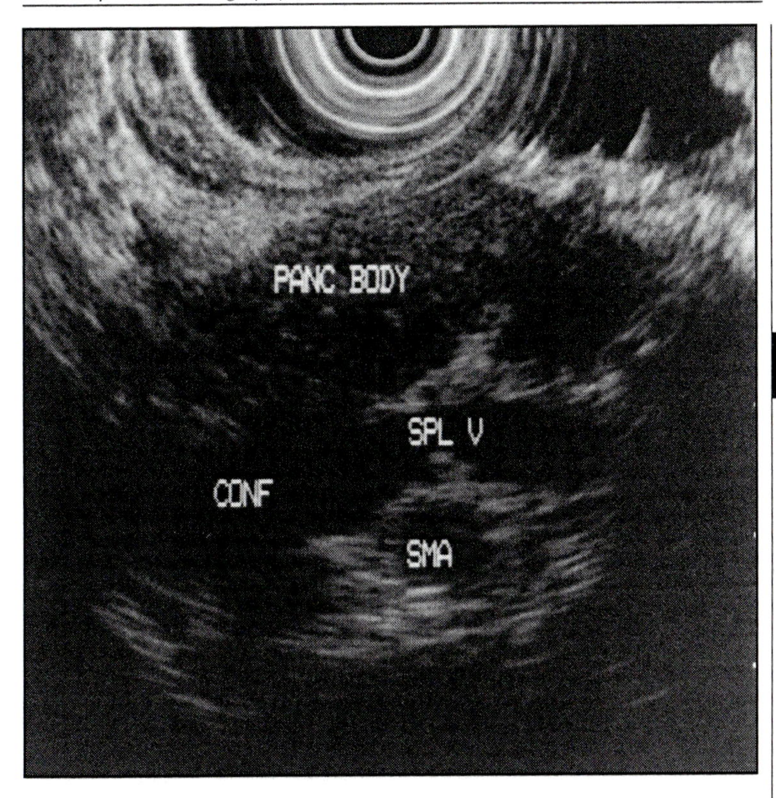

Figure 24.6. Chronic pancreatitis – irregular gland margins. The gland appears lobular. Note the hyperechoic band if tissue indenting the gland just above "SPL V". Contrast this with Figure 2. (CONF = portal confluence, SPL V = splenic vein, SMA = superior mesenteric artery). (Imaged at 7.5 MHz via Olympus GFUM130 videoechoendoscope, Olympus America Inc, Melville, NY).

Endosonographic Evaluation of Cholangiocarcinoma

- Background. **Cholangiocarcinoma**, a malignancy arising in the bile ducts, is the second most prevalent primary liver cancer (behind hepatocellular). These are most commonly adenocarcinomas. Tumors may arise at any point along the intrahepatic or extrahepatic biliary tree. Approximately half occur at the hilum of the liver, the so called "Klatskin tumor". Klatskin tumors have an especially poor prognosis, with an overall 5 year survival rate of 1%.
- Histology of the bile duct. The bile duct is essentially a three-layer structure. An inner mucosal layer is surrounded by muscularis propria. The entire duct is then surrounded by a thin layer of connective tissue (serosa).
- TNM classification
  - Limitations of TNM classification
    - TNM classification of cholangiocarcinoma was first proposed in 1987. Since then it has been criticized because it does not account for tumor

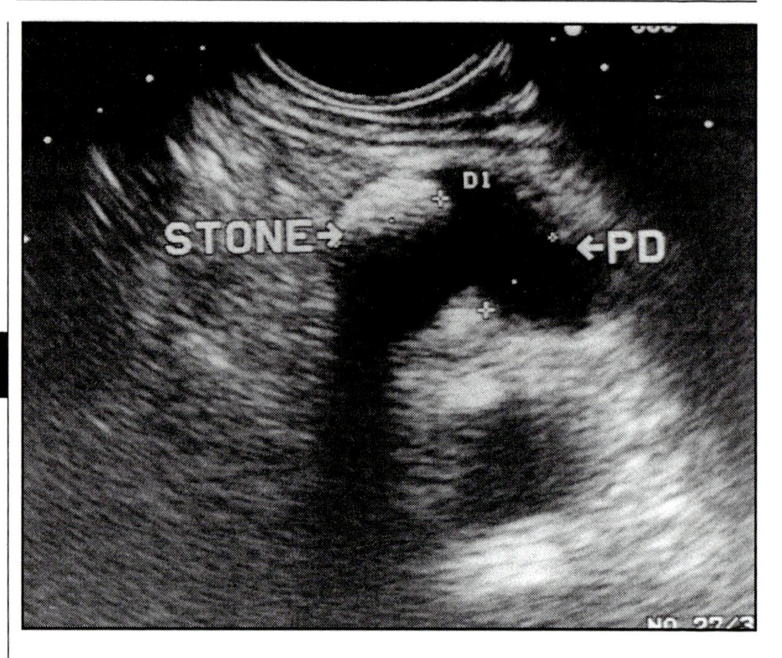

Figure 24.7. Chronic pancreatitis – dilated pancreatic duct with stone. A 9.5 mm stone is seen within a dilated, 8 mm diameter pancreatic duct. Note the acoustic shadowing produced by the stone. Also, note the hyperechoic tissue below the pancreatic duct due to the enhanced through transmission as is seen in cysts (Imaged at 5 MHz via Pentax FG32UA linear array echoendoscope, Pentax Corp, Orangeburg, NY).

location or length of bile duct involvement – these two factors are important prognostically and for determining resectability.

- Primary tumor ("T-stage")
  - Tx Primary tumor cannot be assessed
  - T0 No evidence of primary tumor
  - Tis Carcinoma *in situ*
  - T1 Tumor invades the mucosa or muscle layer
    - T1a Tumor invades the mucosa
    - T1b Tumor invades the muscle layer
  - T2 Tumor invades the perimuscular connective tissue
  - T3 Tumor invades adjacent structures: liver, pancreas, duodenum, gallbladder, colon, stomach
- Regional lymph nodes ("N-stage")
  - See section above for a discussion of EUS characteristics of malignant nodes.
  - Nx Regional lymph nodes cannot be assessed
  - N0 No regional lymph node metastasis
  - N1 Metastasis in the cystic duct, pericholedochal, and/or hilar

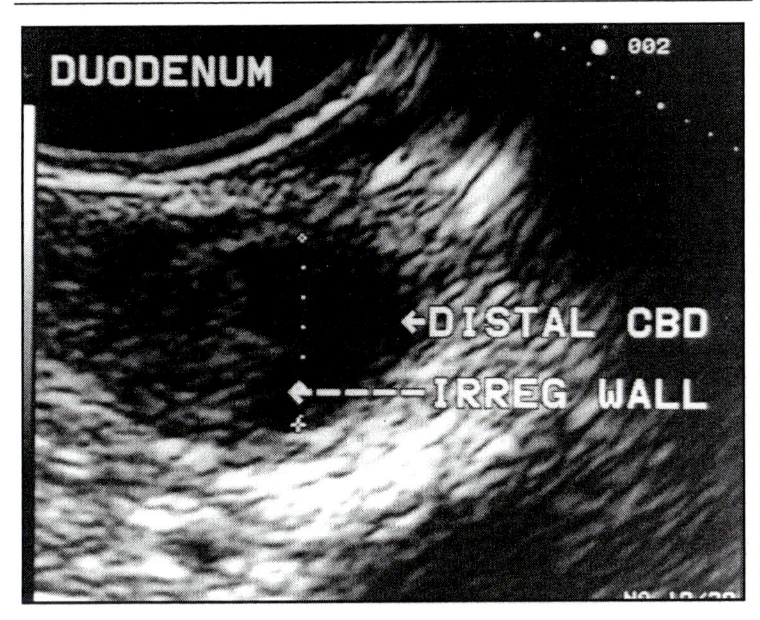

Figure 24.8a. Intraductal tumor of the common bile duct. Linear array cross-sectional image of the distal CBD shows an irregular, semilunar hypoechoic filling defect. (Imaged at 5 MHz via Pentax FG32UA linear array echoendoscope, Pentax Corp, Orangeburg, NY).

lymph nodes
  - N2 Metastasis in peripancreatic (head only), periduodenal, periportal, celiac, superior mesenteric, and/or posterior pancreaticoduodenal lymph nodes
- Distant metastasis ("M-stage")
  - Mx Presence of distant metastasis cannot be assessed
  - M0 No distant metastasis
  - M1 Distant metastasis

• Traditional echoendoscope examination
  - Experience with EUS in the staging of cholangiocarcinoma is limited. Early reports suggest T-staging accuracy of 80-85%. Over-staging may occur due to misinterpretation of fibrosis as tumor. Bile duct wall layers are not easily visualized with traditional EUS.
• Miniprobe examination (Fig. 24.8). Staging accuracy may be enhanced by the use of high-frequency ultrasound probes that can be advanced into the lumen of the bile duct. At 20 MHz, the duct wall is seen as a three-layer structure (hyperechoic mucosa, hypoechoic muscularis, and hyperechoic serosa). Large series have yet to be reported.
• Endosonography of the gallbladder (Fig. 24.9). EUS obtains excellent views of the gallbladder and can readily demonstrate gallstones, sludge and changes of cholecystitis (wall thickening, pericholecystic fluid). For these findings,

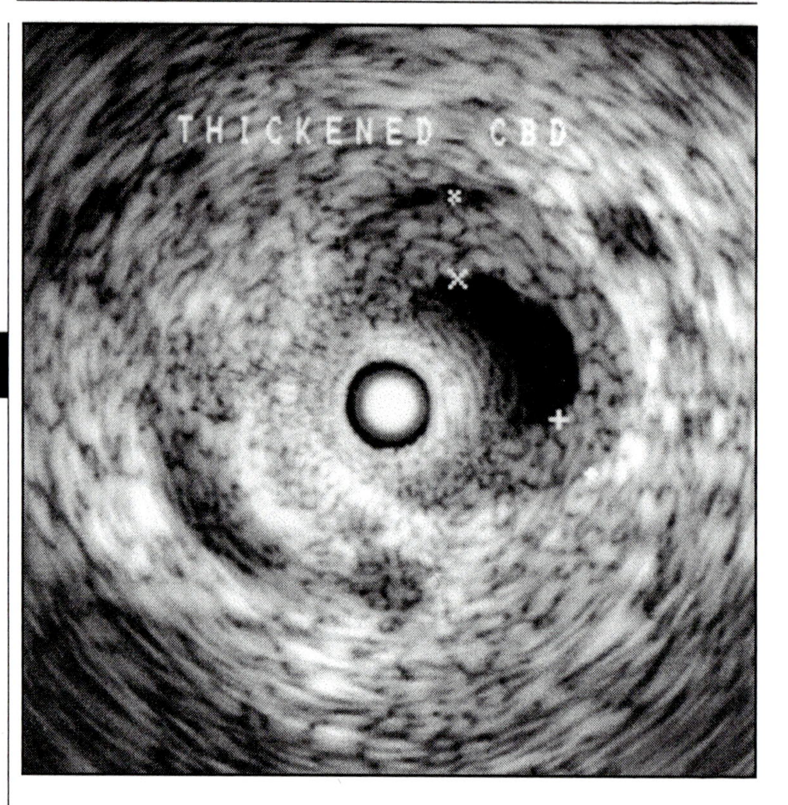

Figure 24.8b. Intraductal tumor of the common bile duct. Intraductal ultrasound of the distal CBD in the same patient shows a thickened wall (measuring 2.2 mm in maximal thickness). The outer surface of the duct is smooth, suggesting a lack of extraductal extension. (Imaged at 12 MHz via Olympus UM2R probe, Olympus America Inc, Melville, NY).

however, there is no documented advantage of EUS over transabdominal ultrasound (which is considerably less invasive). EUS may provide preoperative staging of early gallbladder carcinoma due to its ability to image individual wall layers. It is unclear how this information would alter management.

- Endosonographic staging of ampullary tumors (Fig. 24.10). Tumors of major papilla range from benign tubulovillous adenomas to invasive adenocarcinomas. Forceps biopsy is positive in only about 50% of malignant cases, making preoperative diagnosis difficult. Ampullary tumors occur with high frequency in patients with familial adenomatous polyposis syndrome. In fact, now that prophylactic colectomy is recommended for these patients, ampullary carcinoma represents the most common disease-related cause of mortality and endoscopic screening is recommended.

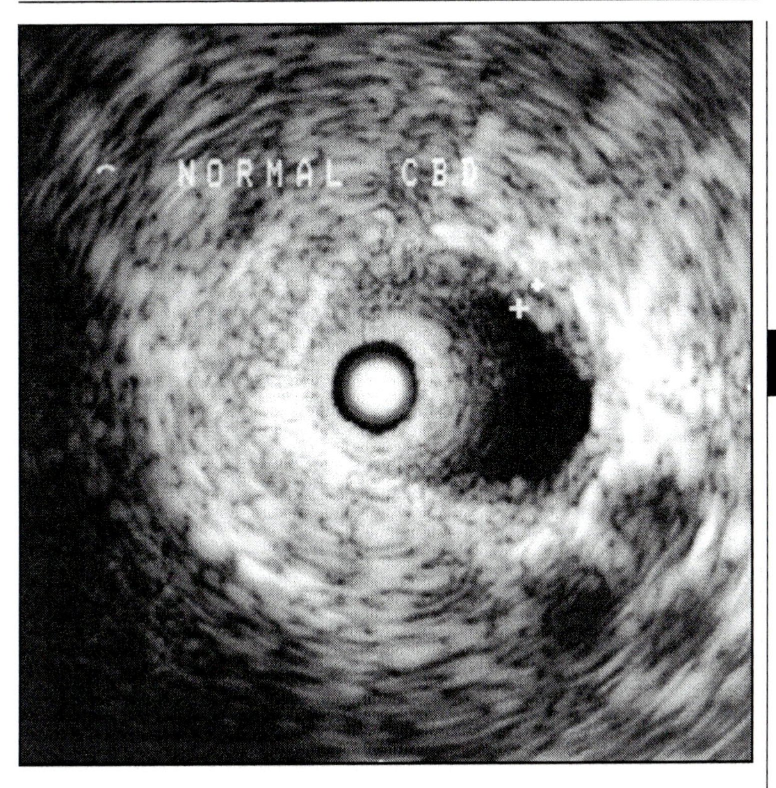

Figure 24.8c. Intraductal tumor of the common bile duct. Intraductal ultrasound of normal bile duct proximal to the lesion seen in 10b. Here the wall measures 0.7 mm in maximal thickness. (Imaged at 12 MHz via Olympus UM2R probe, Olympus America Inc, Melville, NY).

Rationale For Preoperative EUS Staging

- Traditional surgical treatment involves pancreaticoduodenectomy (Whipple procedure). Localized lesions (T1 or T2, N0, M0) can undergo less aggressive localized resection (either via endoscopic snare ampullectomy or surgical ampullectomy) which is considerably less expensive than Whipple. EUS is 80-90% accurate for the identification of advanced disease, and the use of preoperative EUS to select patients for local resection has therefore been suggested to be cost-effective.[8]
  - TNM Staging
    - Primary tumor ("T-stage")
      - Tx Primary tumor cannot be assessed
      - T1 Tumor limited to the ampulla of Vater
      - T2 Tumor invades the duodenal wall

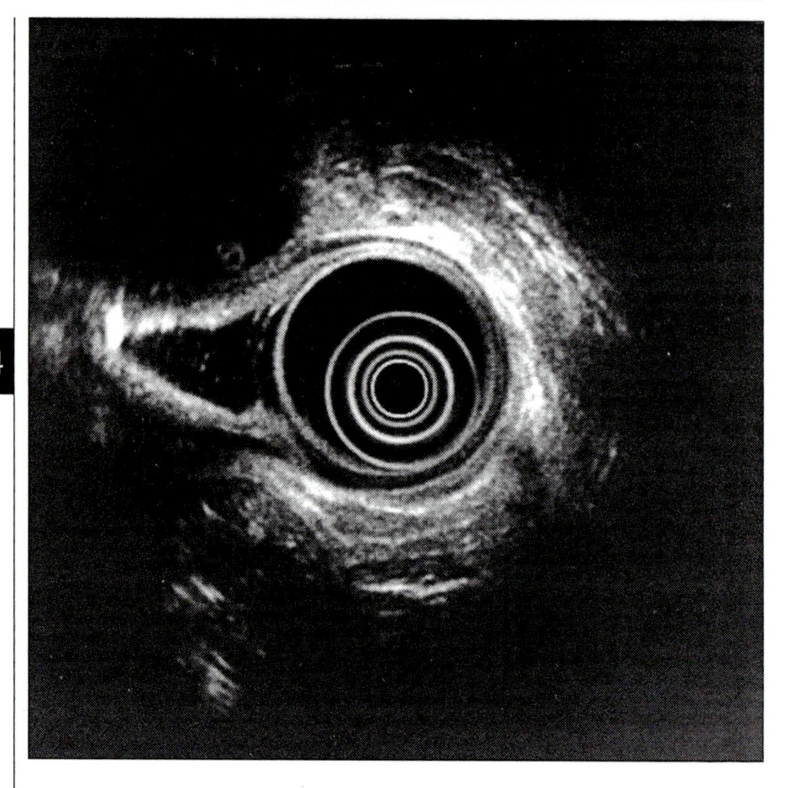

Figure 24.9. Gallbladder polyp. The transducer is placed in the gastric antrum. The semicircular anechoic region between 9 and 11 o'clock is the gallbladder. There is a small (4 mm) round polyp with two internal anechoic regions arising from the wall closest to the transducer. (Imaged at 12 MHz via Olympus GFUM130 videoechoendoscope, Olympus America Inc, Melville, NY).

- T3 Tumor invades £2 cm into the pancreatic parenchyma
- T4 Tumor invading >2 cm into the pancreas and/or adja cent organs

- Regional lymph nodes ("N-stage")
  - See section Above for a discussion of EUS characteristics of malignant nodes.
  - Nx Regional lymph nodes cannot be assessed
  - N0 No regional lymph node metastasis
  - N1 Regional lymph node metastasis
- Distant metastasis ("M-stage")
  - Mx Distant metastasis cannot be assessed
  - M0 No distant metastasis
  - M1 Distant metastasis: Includes lymph nodes along the splenic vein or in the splenic hilum

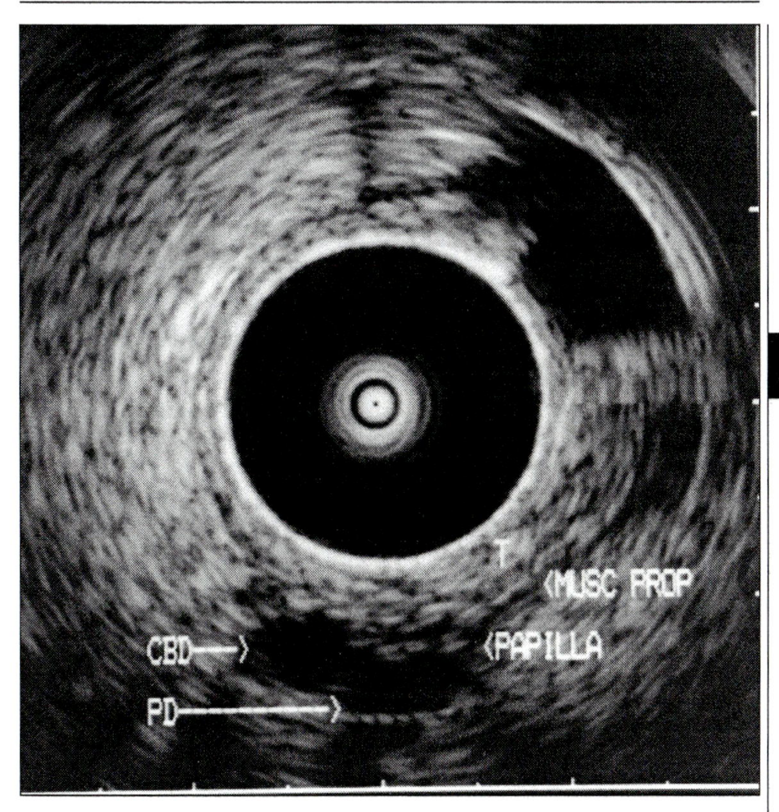

Figure 24.10. Ampullary adenoma. The common bile duct (CBD) and pancreatic duct (PD) enter the papilla at 6 o'clock. The tumor is seen between 3 o'clock and 6 o'clock as an asymmetric thickening of the wall. There is no involvement of the muscularis propria. Snare excision was performed and the lesion was confirmed to be a benign adenoma. (Imaged at 12 MHz via Olympus UM2R probe, Olympus America Inc, Melville, NY).

Miscellaneous Indications for Endosonography

- EUS-guided celiac neurolysis. Many of the nerve signals transmitting pain from the pancreas pass through the celiac plexus. Percutaneous celiac nerve block has been shown to be effective in improving the pain associated with pancreatic malignancies. Recent studies have shown that EUS can be used to inject local anesthetic agents or absolute alcohol into the region of the celiac ganglion with similar results.[9] Although the celiac ganglion itself cannot be visualized, its proximity to the celiac trunk has been well described. The advantage of EUS-guided nerve block is that neurolysis can be performed at the same time as initial EUS diagnosis and staging.

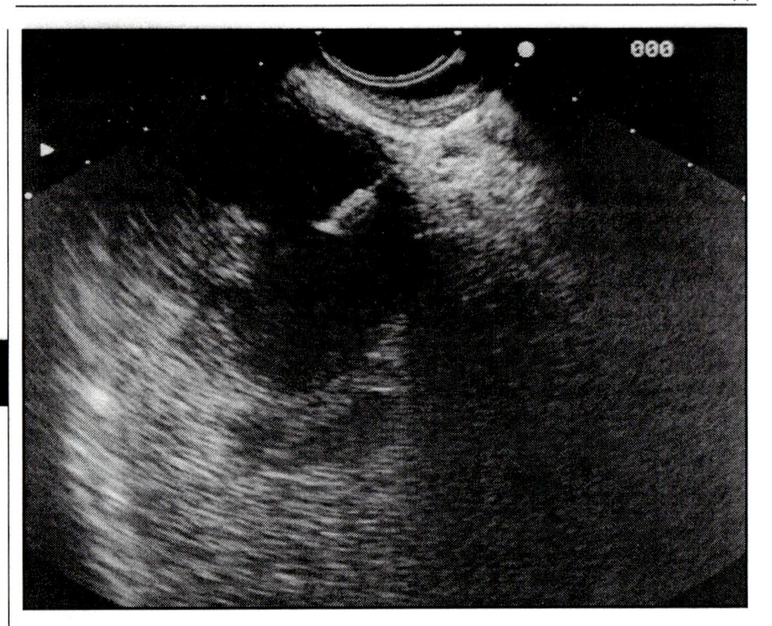

Figure 24.11. EUS-guided needle aspiration of a splenic abscess. An 86 year old man was found to have a necrotic adenocarcinoma of the pancreatic tail. An incidental abnormality was seen in the spleen, which is seen filling the left half of the image. There is an irregular, 3 x 5 cm hypoechoic region present in the superior pole (top of image). Fine needle aspiration was performed to ruleout metastasis. A 22 gauge needle can be seen entering the spleen from the upper right, with the tip centered in the cavity. Aspiration returned pus and cultures grew gram negative rods. (Imaged at 5 MHz via Pentax FG32UA linear array echoendoscope, Pentax Corp, Orangeburg, NY).

EUS of the Spleen (Fig. 24.11)

- Intrasplenic lesions may be seen due to metastatic disease, embolic disease, granulomatous disease or rarely pyogenic abscess. Typically these are diagnosed and managed via other imaging modalities; however occasionally unsuspected lesions are identified at the time of diagnostic EUS.

EUS of the Upper Abdominal Vasculature (Fig. 24.12)

- Several miscellaneous disorders of the upper abdominal vasculature are potentially amenable to diagnosis via EUS; however the utility of EUS for these disorders has not been extensively evaluated. Vascular disorders for which EUS may prove useful include mesenteric vascular insufficiency and splenic vein thrombosis.

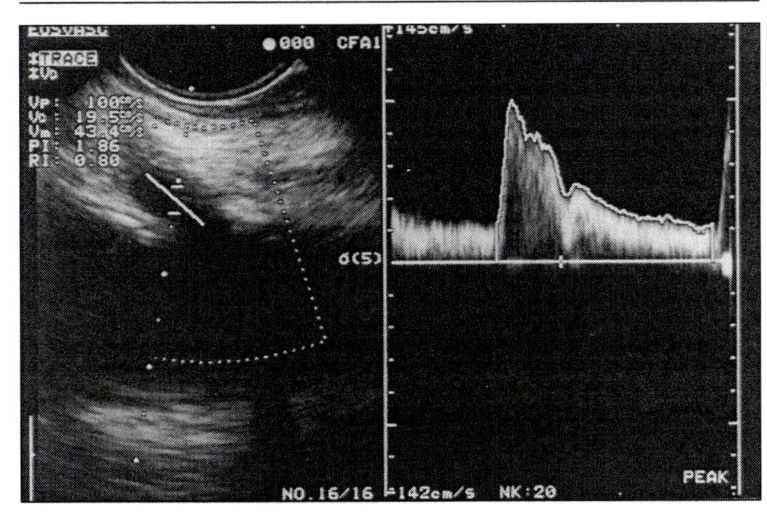

Figure 24.12. Doppler evaluation of the celiac trunk. In the left image, the aorta is seen as an anechoic horizontal band across the middle of the image. A cursor is placed within the celiac trunk. The graph on the right displays flow velocity as a function of time. Note: a second branch can be seen parallel to the celiac trunk at the left edge of the image—this is the superior mesenteric artery. (Imaged at 5 MHz via Pentax FG32UA linear array echoendoscope, Pentax Corp, Orangeburg, NY).

Future Applications of EUS

- The ability to direct the placement of needles within focal lesions via EUS has resulted in a dramatic shift in the utility of EUS for gastrointestinal malignancies. EUS has progressed from being a simple imaging study to a study that allows directed tissue sampling. Many endosonographers foresee a time when therapeutic intervention will also be possible. Initial studies have suggested that EUS can be used to inject therapeutic substances directly into pancreatic tumors or apply thermal therapy for the focal ablation of neoplastic tissue.
- EUS-guided fine-needle injection (FNI). Preliminary data regarding the injection of activated T-lymphocytes directly into pancreatic tumors was recently reported in abstract form. Other agents of interest include chemotherapeutic agents and viral vectors for gene therapy.
- EUS-guided thermal ablation. Several studies have recently demonstrated the clinical promise of percutaneous needle-guided thermal ablation of small neoplasms. This can involve extreme cold (cryoablation) or heat (radio-frequency ablation). A preliminary report demonstrated the technical feasibility of performing EUS-guided radio-frequency ablation in a porcine model of the human pancreas. This could potentially be useful for the treatment of functional neuroendocrine tumors, tumors of borderline malignant potential or in nonsurgical candidates.

Selected References

1. Catalano MF, Sivak MV Jr, Rice T et al. Endosonographic features predictive of lymph node metastasis. Gastrointest Endosc 1994; 40:442-446.
2. Müller MF, Meyenberger C, Bertschinger P et al. Pancreatic tumors: Evaluation with endoscopic US, CT and MR imaging. Radiology 1994; 190:745-751.
3. Chang KJ, Nguyen P, Erickson RA et al. The clinical utility of endoscopic ultrasound-guided fine-needle aspiration in the diagnosis and staging of pancreatic carcinoma. Gastrointest Endosc 1997; 45:387-393.
4. Rösch T, Braig C, Gain T et al. Staging pancreatic and ampullary carcinoma by endoscopic ultrasonography. Gastroenterology 1992; 102:188-199.
5. Rösch T, Lightdale CJ, Botet JF et al. Localization of pancreatic endocrine tumors by endoscopic ultrasonography. N Engl J Med 1992; 326:1721-1726.
6. Sahai AV, Zimmerman M, Aabakken L et al. Prospective assessment of the ability of endoscopic ultrasound to diagnose, exclude, or establish the severity of chronic pancreatitis found by endoscopic retrograde cholangiopancreatography. Gastrointest Endosc 1998; 48:18-25.
7. Canto MIF, Chak A, Stellato T et al. Endoscopic ultrasonography versus cholangiography for the diagnosis of choledocholithiasis. Gastrointest Endosc 1998; 47:439-448.
8. Quirk DM, Rattner DW, Fernandezdel Castillo C et al. The use of endoscopic ultrasonography to reduce the cost of treating ampullary tumors. Gastrointest Endosc 1997; 46:334-337.
9. Wiersema MJ, Wiersema LM. Endosonography-guided celiac plexus neurolysis. Gastrointest Endosc 1996; 44:656-662.

CHAPTER 25

Liver Biopsy

David Bernstein

Introduction

Needle biopsy of the liver, when used appropriately, is a vital diagnostic modality in the evaluation of hepatobiliary disorders. Percutaneous needle biopsy can be performed by gastroenterologists or hepatologists with or without the use of ultrasound guidance. Various techniques of percutaneous liver biopsy have been described and the choice of technique should be made based upon the clinical circumstances and the comfort level of the person performing the biopsy. Radiologists, surgeons and laparoscopists also perform liver biopsies. Regardless of technique, liver biopsy is generally safe with an overall complication rate ranging from 0.06-0.32% and a mortality rate of less than 0.01%.

Relevant Anatomy

- Thorough understanding of the anatomy of the liver and its surroundings is essential when performing a liver biopsy. This information may help prevent complications as well as provide an understanding as to why an untoward event may occur.
    - The liver is the largest solid organ in the body lying largely under the right lower rib cage. The normal liver span space in the mid-clavicular line is from the fifth intercostal to the right costal margin. Because of this location, it is usually not palpable on physical examination. Therefore, careful percussion is required to accurately determine its location and size. The liver is covered by Glisson's capsule and divided into four lobes: right, left, caudate and quadrate. (Fig. 25.1)
    - Both superiorly and laterally, the liver is attached to the diaphragm. Anteriorly, the liver is connected to the anterior abdominal wall and diaphragm by the falciform ligament. This ligament, which contains the remains of the umbilical vein, is called the round ligament in its lower end. At its upper end, the falciform ligament ascends the anterior surface of the liver and divides the liver into the left and right lobes. The posterior surface of the right lobe abuts the colon, right kidney and duodenum from the right while the posterior left lobe abuts the stomach. (Fig. 25.2) Reidel's lobe is a benign anatomic variant resulting in an enlarged right lobe that propagates caudally and can be easily confused with pathologic hepatomegaly. The gallbladder lies against the inferior surface of the right and quadrate lobes.
    - Percutaneous liver biopsy is performed by passing a needle through an intercostal space. The most common space utilized is the interspace between the 8th and 9th ribs. Intercostal nerves, muscles and vessels run make up the intercostal space. The anterior intercostal arteries and veins and the intercostal nerves run along the lower border of each ribs. Percutaneous liver

Gastrointestinal Endoscopy, edited by Jacques Van Dam and Richard C. K. Wong.

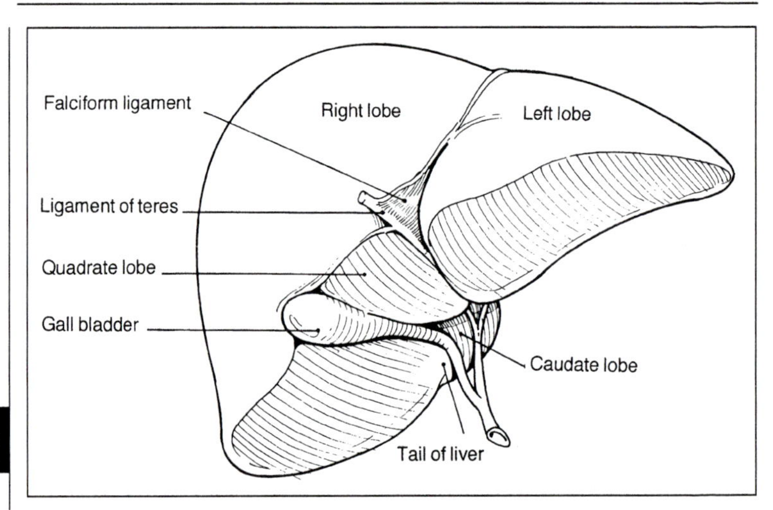

Figure 25.1. Anatomy of the liver.

biopsy is thus performed by passing the biopsy needle over the superior aspect of the rib to avoid puncturing the intercostal vessels and nerves.

Indications and Contraindications

- Indications for percutaneous biopsy of the right lobe of the liver
  - Evaluation of abnormal liver enzymes of greater then six months duration
  - Staging of chronic liver disease (ex. alcoholic liver disease, chronic hepatitis B and C, primary biliary cirrhosis, autoimmune hepatitis, etc.) to determine the extent of inflammation, fibrosis and cirrhosis.
  - Evaluation of unexplained jaundice
  - Evaluation of unexplained hepatomegaly
  - Evaluation of fever of unknown origin. Multiple passes should be obtained to increase the yield of diagnosis. The biopsy specimen should be sent for routine pathology, viral studies and bacterial and fungal culture.
  - Evaluation of acute and chronic drug-induced injury
  - Evaluation of intrahepatic masses utilizing either ultrasound or CT guidance
  - Evaluation of patients with hereditary liver disease such as hemochromatosis, Wilson's disease, alpha1 antitrypsin deficiency, glycogen storage diseases and tyrosinemia to determine extent of disease and assess response to therapy.
  - Evaluation of patients receiving more than 1.5-2 g of methotrexate therapy for psoriasis or rheumatoid arthritis. These patients may require multiple liver biopsies over their course of treatment.
  - Recognition of systemic or granulomatous disorders.
  - Evaluation of status following liver transplantation to determine the presence of rejection, recurrent disease, cytomegalovirus infection, and ischemic liver disease.

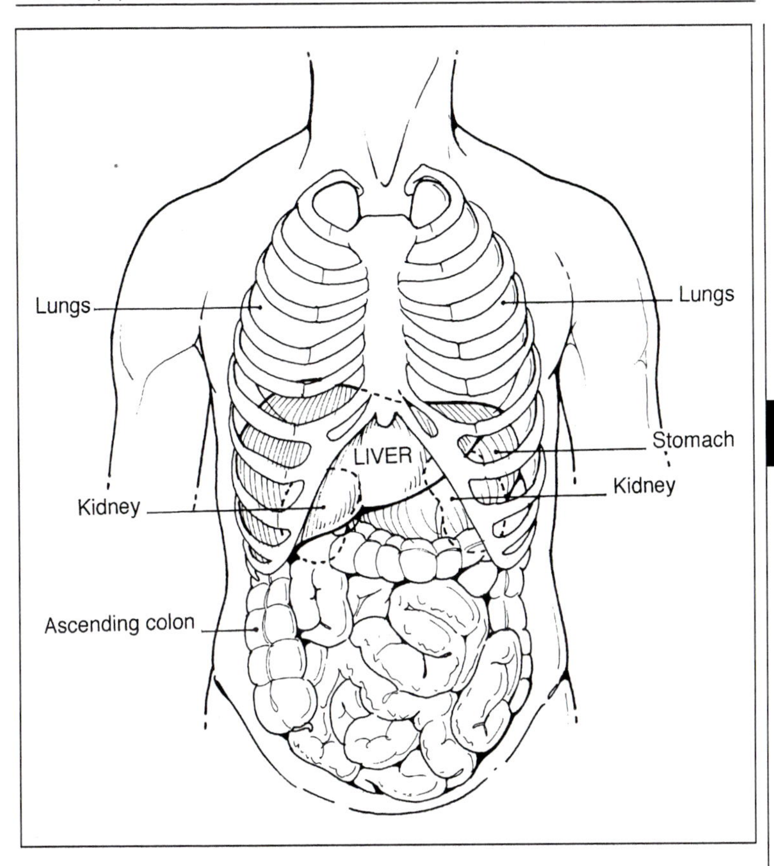

Figure 25.2. The liver and its surrounding organs.

- Contraindications
  - Absolute
    - Uncooperative patient
    - Coagulopathy
    - Prothrombin time > 3 seconds above control. INR can not be used in the setting of advanced liver disease secondary to the inconsistent and variable coagulation factor levels seen in liver disease.
    - Platelet count less than 80,000/mm^3
    - Prolonged bleeding time
    - Unavailability of blood transfusion support
    - Ascites
    - Presence of vascular tumor
    - Serious consideration of echinococcal disease
    - Polycystic liver disease

- Relative
  - Infections in the right pleural cavity
  - Infection below the right diaphragm
- Laparoscopic liver biopsy
  - Biopsies are usually obtained from the left lobe although biopsies of the right lobe can also be obtained.
  - Indications: In addition to all the indications listed above for percutaneous liver biopsy, laparoscopic liver biopsy may be indicated for:
    - Ascites of unknown etiology
    - Staging of malignancy such as pancreatic, cholangiocarcinoma, esophageal, and gastric.
    - Evaluation of peritoneal infections
  - Contraindications
    - Uncooperative patient
    - Coagulopathy as with percutaneous liver biopsy
    - Severe cardiopulmonary failure
    - Abdominal wall infection
    - Intestinal obstruction
    - Irreducible external hernia

Equipment

- Percutaneous needle biopsy selection
  - Suction needles: these needles obtain a core biopsy of liver tissue with the aid of suction and a specially designed, blunt ended needle. The cannula is 6 cm long with a needle wall diameter of 90 μm. The needle does not have a stylet but is equipped with a stopping device to prevent the tissue specimen from being sucked into the syringe providing suction. The suction needle technique is very simple and safe as the time the needle is in the liver parenchyma for only 0.51 seconds. These needles, however, are associated with greater specimen fragmentation.
    - Reusable
      - Menghini: this needle is rarely used and has been largely replaced by the Jamshidi needle
      - Klatskin needle
    - Disposable
      - Jamshidi
- Cutting needles: these needles obtain a core biopsy utilizing a cutting technique. The most common needles are Trucut needles. Newer, spring-loaded devices have been introduced which decrease the time that the needle is within the liver to microseconds. These needles have less fragmentation and are recommended for use in cirrhotic patients.
  - Types of cutting needles:
    - Reusable
      - VimSilverman: this has been replaced by newer needles as the time the needle is in the liver, usually 5-10 ses, is associated with more frequent complications.
    - Disposable
      - Trucut
      - Spring-loaded devices

Table 25.1. Criteria for outpatient liver biopsy

- Patient must be able to easily return to the hospital where procedure was performed within 30 minutes of any untoward symptoms.
- Patient must have a reliable individual stay with him/her during the first post-biopsy night to provide care and transportation if necessary.
- Patient should have no complications or associated serious medical problems that increase the risk of biopsy.
- Facility where the biopsy is performed should have an approved laboratory, blood-banking unit, easy access to an inpatient bed and personnel to monitor spitalized after the biopsy if any serious or persistent complications develop.

American College of Physicians Guidelines, Clinical Competence in Liver biopsy

Procedure

- Percutaneous liver biopsy. Outpatient biopsies should be performed before 12 PM in either the patient's room or the outpatient unit.
  - Preprocedure assessment. More than 95% of liver biopsies performed in the United States are done in the outpatient setting. Criteria for outpatient liver biopsy are seen in Table 25.1. Therefore, the biopsy procedure should be explained to the patient prior to the biopsy date to avoid misunderstandings and increase patient awareness. Despite this, most patients are anxious and nervous on the day of their liver biopsy. A confident, calm and reassuring demeanor is essential in allowing patients to relax and feel comfortable. Therefore, it is important to gather as much information as possible about the patient prior to the liver biopsy.
  - History. A thorough history including past medical, past surgical, social and medication history is essential. Hemophilia and renal failure are associated with an increased risk of bleeding complications. Patients with arthritis, bursitis or other musculoskeletal disorders are more likely to have difficulty lying on their right side and back for the six hours following the liver biopsy. Patients on nonsteroidal antiinflammatory medications, antiplatelet medications, and anticoagulants will also require special management.
  - Physical examination. Baseline blood pressure, pulse and respiratory rate are obtained. Special attention should be given to the lung, cardiac and abdominal exams. Diaphragmatic excursion should be assessed. Percussion evaluates the liver span and the liver edge examined. Surgical scars should be noted.
  - Laboratory analysis. Routine laboratory tests prior to liver biopsy include: complete blood count with platelet count, prothrombin time, activated partial thromboplastin time, type and screen, serum creatinine level and ultrasound of liver. Serum tests should be obtained within a seventy-two hours of the liver biopsy. Routine bleeding times have not been shown to be of value in predicting bleeding complications. Therefore, a bleeding time should only be obtained in patients with low platelets, abnormal coagulation studies, jaundice, poor hepatic function, underlying hematologic disorders or renal disease.
  - Coagulation status. Patient on nonsteroidals or antiplatelet medications: these medications should be stopped one week prior to and not restarted for

two weeks after performance of the liver biopsy. This may require consultation with the patient's primary care physician, cardiologist, neurologist or rheumatologist.

- Heparinized patient. Heparin should be stopped 48 hours prior to the procedure to allow the aPTT to normalize. Twenty-four hours following the biopsy, heparin can be restarted.
- Warfarin should be discontinued at least 72 hours prior to the biopsy and a prothrombin time checked immediately before the biopsy. Warfarin can be started 48-72 hours after the biopsy is complete. If heparin is needed, it can be started 24 hours after the biopsy. It is not advisable to start warfarin immediately after the biopsy despite its half-life of 72-96 hours as delayed bleeding has been reported.
- Special circumstances
  - Thrombocytopenia. Platelet transfusions can be given to keep the platelet count above 80,000/mm^3. These patients should be observed overnight in the hospital following liver biopsy.
  - Prolonged prothrombin time > 3 seconds above control: Fresh frozen plasma can be given to improve the prothrombin time to less than 3 sec above control. These patients should be observed overnight in the hospital following liver biopsy.
  - Chronic renal failure. Patients on hemodialysis should undergo biopsy the day after dialysis. Regardless of clotting study results, patients should be given a single dose of desaminoDarginine vasopressin (DDAVP). The dose of heparin given on the first session of dialysis after biopsy should be reduced. These patients should be observed overnight in the hospital following liver biopsy.
  - Hemophiliacs can safely undergo liver biopsy with adequate factor VIII or IX replacement. Consultation with a hematologist is essential. These patients should be observed overnight in the hospital following liver biopsy.
  - Well established cirrhosis or tumors. Liver biopsies should not be performed on an outpatient basis since fatal complications such as hemoperitoneum have been associated with these conditions.
  - AIDS. The literature is inconclusive if patients with human immunodeficiency virus infection are at increased risk of bleeding following liver biopsy. When indicated and in the absence of contraindications, liver biopsy should be performed.
- Special preprocedure instructions. Patients should eat a light breakfast including milk or another dairy product to stimulate gallbladder contraction. Someone must accompany the patient to and from the outpatient biopsy setting
- Obtain informed consent
- Obtain intravenous access
- It is not necessary to premedicate the patient with either meperidine or midazolam although these can be used in the anxious patient.
- Ask patient to urinate or defecate if needed
- Place patient in supine position near the side of the bed or stretcher
- Right hand is placed under head
- Left arm is placed along side the left side, parallel to the body.

- Biopsy site is determined by percussion over the right hemithorax in the midaxillary line until an area of maximum dullness is identified between the ribs at the end of expiration. The biopsy site is selected at one interspace below this point. If ultrasound is available, it can be used at this point to confirm the biopsy site.
- Skin around and over the biopsy site is cleaned with antiseptic.
- The biopsy site is draped in sterile fashion
- A local anesthetic, preferably 1% lidocaine without epinephrine, is infiltrated into the skin to produce a small bleb. Approximately 45 cc of local anesthetic is given in the intercostal space, above the rib. Further anesthesia is given down to the level of the liver capsule and peritoneum.
- The skin is punctured with a scalpel and the biopsy needle of choice is inserted through the incision.
- Suction technique
  - Approximately 5 cc NS is loaded into the syringe and the biopsy needle is flushed to remove all air.
  - The biopsy needle is inserted approximately 5 mm through the internal fascia of the intercostal muscles until a popping sensation is felt.
  - Approximately 12 cc of saline is instilled to clear the needle of debris.
  - Constant suction is applied on the plunger by pulling the syringe back three notches and locking it in place.
  - The patient is asked to take a deep breath, exhale and hold his breath in the end of expiration.
  - The needle is rapidly inserted and removed from the liver and withdrawn from the skin.
  - The biopsy specimen is then removed by slowly flushing the needle with saline over a sterile container or gauze.
- Cutting technique (non-automated needle)
  - The needle is inserted into the puncture site and advanced through the intercostal muscles.
  - The patient is asked to take a deep breath, exhale and hold the breath at the end of expiration.
  - The needle is inserted 23 cm within the liver in a downward direction.
  - The inner trocar is advanced holding the outer sheath steady.
  - The outer cutting sheath is advanced to cut the liver.
  - The entire needle is removed.
- Cutting technique (automated needle)
  - The needle is readied, inserted into the puncture site and advanced through the intercostal muscles.
  - The patient is asked to take a deep breath, exhale and hold the breath at the end of expiration.
  - The needle is inserted 23 cm within the liver in a downward direction.
  - The needle is discharged and removed.
- Post-biopsy procedure and orders
  - Bandage is placed over puncture site.
  - Patient is placed in right lateral decubitus position for 3 h then placed supine for 3 additional hours.
  - Vital signs are measured every 15 min for the first 3 h then every 30 min for the following 3 h.

Table 25.2. Common pathologic stains used to evaluate the liver

Stain	Usage
Hematoxylin and eosin	Routine evaluation
Perl's Prussian Blue	Iron Content
Diastase/PAS	Alpha1 antitrypsin deficiency
Trichrome	Fibrosis
Reticulin	Fibrosis
Orcein or Victoria blue stain	Elastic fibers, copper-associated protein, Hepatitis B
Rhodanine	Copper
Congo Red	Amyloid

- Diet may be resumed within 2 h of the biopsy.
- Analgesia such as acetaminophen or oxycodone plus acetaminophen can be given if needed.
- Patient may not get up to use the bathroom for 6 h.
- Notify physician on call if blood pressure < 90/60, pulse > 110 beats per minute or severe pain.
- After 6 h, patient may be discharged home.

• Laparoscopic liver biopsy
  - Technique
    - Obtain informed consent
    - Obtain intravenous access
    - Place in supine position
    - Premedicate with intramuscular meperidine
    - Abdomen draped and prepped in sterile fashion.
    - Intravenous meperidine and midazolam given for conscious sedation
    - An area of the abdominal wall approximately 2 cm above and 2 cm to the left of the umbilicus is anesthetized with 1% lidocaine without epinephrine.
    - Insert Veress needle and achieve pneumoperitoneum with either nitrous oxide or carbon dioxide.
    - Remove Veress needle.
    - Insert trocar into abdominal wall through Veress needle site and remove needle.
    - Insert laparoscope through trocar.
    - Place patient in reverse Trendelenburg position.
    - Perform visual exam with a laparoscope.
    - A biopsy needle is introduced through the abdominal wall under laparoscopic guidance and advanced to the liver in a tangential approach.
    - The needle is introduced into the liver and the biopsy is performed.
    - Insert second trocar through the abdominal wall to facilitate tamponade of biopsy site. Observe biopsy site until hemostasis occurs.
    - Remove both trocars with careful observation of sites for bleeding.
    - Suture puncture wounds after pneumperitoneum is allowed to drain.
    - Place bandage over puncture site
    - Observe patient overnight in hospital

Outcome

- An adequate sample should be at least 1.5 cm in length with at least five portal areas visible on the specimen. Cholestasis, steatosis, and hepatitis are relatively diffuse processes and smaller specimens may provide the correct diagnosis. Cirrhosis, however, is more difficult to diagnosis by blind needle biopsy. The false negative rate for cirrhosis on specimens obtained from percutaneous liver biopsy is approximately 30%. Laparoscopic liver biopsy has been shown to decrease this false negative rate to less than 5%.
- After the biopsy is performed, processing of the sample becomes of paramount importance. The specimen must be divided before being placed in formalin solution if special stains are required. Unfixed tissue is required for viral and mycobacterial cultures. Fixation in 3% buffered glutaraldehyde is needed for electron microscopy. Unfixed tissue must be placed in metal free containers if analysis for iron or copper is warranted. Rapid freezing in liquid nitrogen or dry ice is needed for immunohistochemical and enzyme activity studies, quantitative hormone receptor studies and isolation of genomic and viral DNA and RNA for molecular analysis. All biopsy specimens should be sent for the following stains: hemotoxylin and eosin, iron, diastase/PAS and Masson's trichrome. (Table 25.2)

Complications

- **Pain** is the most common complication following liver biopsy occurring in up to half of patients. It usually begins approximately 10-15 min after the biopsy is complete and coincides with the loss of local analgesic effect given during to the procedure. Pain commonly occurs over the biopsy site, at the right shoulder and is often described as a drawing feeling across the epigastrium. Most of these episodes of pain are self-limited and resolve within the first 12 hours. However, 10% of all patients will require pain relief with either acetaminophen with or without oxycodone.
- **Hemorrhage** is the most worrisome complication of liver biopsy. The incidence of clinically significant bleeding resulting in intraperitoneal hemorrhage is 0.03 – 0.7 %. Most bleeding occurs within the first 23 hours with 95% of bleeding episodes occurring within the first six hours after liver biopsy. Significant bleeding may be secondary to liver laceration occurring at the time of expiration or from the puncture of the hepatic or portal vein. Risk factors for bleeding include malignancy, advanced age, increased number of passes with the biopsy needle, and the presence of coagulopathy. Clinical signs of significant hemorrhage, hypotension, tachycardia and severe abdominal pain necessitate immediate action. A small percentage of these bleeds will respond to medical therapy; however, most will require either radiological or surgical intervention. Transfusion of blood and blood products is usually required. A bedside ultrasound may demonstrate a free intraperitoneal collection not previously present. Angiography is the preferred treatment. If this is not available or is unable to stop the bleeding, urgent surgery must be performed.
- **Pleurisy** may occur after liver biopsy and usually resolves on its own. A friction rub may develop and last for weeks following a biopsy.
- **Hemothorax** may occur with injury to an intercostal vessel. This can be avoided by passage the needle just above the rib.
- **Pneumothorax** is usually mild with pulmonary collapse generally less than 10% of lung volume. Conservative treatment is effective.

- **Intrahepatic hematomas** are usually asymptomatic and occur in 23% of patients in the 24 hours following a biopsy. Rarely, these can cause pain, fever, elevated transaminases, decreased hematocrit and hemodynamic instability.
- **Hemobilia** usually presents with gastrointestinal bleeding, biliary colic and jaundice. Rarely, it can lead to cholecystitis and pancreatitis. Most of the time bleeding is arterial and patients usually present with symptoms 5-10 days following the biopsy. Biliary scintigraphy, ultrasound and ERCP are helpful in diagnosis. Most cases resolve spontaneously. Endoscopic sphincterotomy will facilitate clot removal in symptomatic patients. If bleeding continues, angiography with embolization is the treatment of choice.
- **Biliary peritonitis** results from the rupture of dilated bile ducts or inadvertent puncture of the gallbladder and is most common when biopsy is performed in the presence of obstructive jaundice. When utilizing the suction technique, bile is seen in the syringe upon withdrawal. Severe pain and hypotension follow the biopsy. This is usually self-limited and responds to analgesia. Biliary scintigraphy and ERCP can localize the leak and the establishment of biliary draining, either endoscopically or transhepatically, allows of closure of the leak.
- Arteriovenous fistulae usually close spontaneously. Angiography with embolization is the treatment of choice should bleeding persist.
- Puncture of an adjacent organ such as the colon, kidney, lung and small bowel is uncommon and rarely clinically significant.
- Bacteremia is relatively common occurring in 13.5% of patients. Sepsis rarely occurs and when present, *E. coli* is the most common organism.
- Vasovagal reactions are common in anxious patients. Reassurance and anti-anxiolytics are helpful in this situation.
- Death is reported in large biopsy series to occur in 0.009-0.12% of cases following liver biopsy.
- **Laparoscopic liver biopsy: unique complications**
  - Cardiac arrhythmia
  - Gas embolism
  - Cerebrovascular accident
  - Pneumoomentum
  - Wound infection
  - Ascites fluid leak
  - Subcutaneous emphysema
  - Abdominal wall hematoma
  - Massive bleeding from lacerated abdominal wall vessel

25

Selected References

1. Ewe K. Bleeding after liver biopsy does not correlate with indices of peripheral coagulation. Dig Dis Sci 1981; 26:388-393.
2. McClosky RV, Gold M, Weser E. Bacteremia after liver biopsy. Arch Intern Med 1973; 132:213-215.
3. McVay PA, Toy PT. Lack of increased bleeding after liver biopsy in patients with mild hemostatic abnormalities. Am J Clin Pathol 1990; 94:747-753.
4. McGill DB, Rakela J, Zinsmeister AR. A 21 year experience with major hemorrhage after percutaneous liver biopsy. Gastroenterology 1990; 99:1396-1400.
5. Piccinino F, Sagnelli E, Pasquale G et al. Complications following percutaneous liver biopsy: A multicenter retrospective study on 68,276 biopsies. J Hepatology 1986; 2:165-173.

6. Sandblum P. Hemobilia. In: Schiff L, Schiff ER, eds. Diseases of the Liver. 7th ed. Philadelphia: J.B. Lippincott Company, 1993:1489-1497.
7. Schiff ER, Schiff L. Needle Biopsy of the liver. In: Schiff L, Schiff ER, eds. Diseases of the Liver. 7th ed. Philadelphia: J.B. Lippincott Company, 1993:216-225.
8. Terry R. Risks of needle biopsy of the liver. Brit Med J 1952; 1102-1105.
9. Thung SN, Schaffner F. Liver biopsy. In: Macsween RNM, ed. Pathology of the Liver. 3rd ed. New York: Churchill Livingstone, 1994; 787-796.
10. Vargas C, Bernstein DE, Reddy KR et al. Diagnostic laparoscopy: A 5 year experience in a hepatology training program. Amer J of Gastro 1995; 90:1258-1262.

CHAPTER 26

Endoscopy of the Pediatric Patient

Victor L. Fox

Introduction

Endoscopy of the pediatric patient differs in many respects from that of the adult patient. Indications for endoscopy in children cover a broad spectrum of acute and chronic conditions, more commonly including congenital than neoplastic disease. Gastrointestinal anatomy and physiology of the developing child must be considered when utilizing drugs for sedation, monitoring cardiovascular and respiratory signs and choosing appropriate endoscopes and accessories for small or delicate structures. Procedural techniques must be adapted to the age and size of the child. Endoscopy of the pediatric patient is optimal when conducted by experienced staff with adequate technical and cognitive skills in a facility that can support the unique physical and emotional needs of a child.

Anatomy and Physiology

- General considerations
  - The **luminal wall** is thinner is infants and young children, increasing the risk of transmural injury or perforation during passage of the endoscope around tight bends or while applying thermal or other tissue destructive techniques.
  - **Cardiovascular and respiratory function** differs dramatically in neonates and young infants.
    - Total blood volume is quite limited—60 to 80 cc/kg body weight. Therefore, relatively small amounts of blood loss can have profound effects.
    - Early recognition of hypovolemic shock is difficult in an infant or young child. Cardiac output in the infant and young child depends more on heart rate than stroke volume. Early signs of volume depletion include tachycardia, decreased capillary refill, and unexplained irritability. Hypotension is often a late sign.
    - Increased vagal nerve response may easily trigger bradycardia.
    - The **tracheal cartilage** is soft (tracheomalacia) and may be compressed by an endoscope or by excessive neck flexion.
    - Chest wall and diaphragm muscles are relative weak. Respiratory function may be compromised by excessive distension of the stomach or intestine with air.
  - Thermoregulation
    - **Temperature regulation** in infants is labile due to increased ratio of body surface area to mass. Hypothermia may occur rapidly and exacerbate pulmonary vascular constriction and acidosis thereby reducing oxygen exchange.

Gastrointestinal Endoscopy, edited by Jacques Van Dam and Richard C. K. Wong.

- Pharmacology. Drug pharmacokinetics and metabolism in children necessitate relatively larger and more frequent doses than in adults. Doses must be calculated on the basis of body weight.
- Behavior. Behavioral responses to parental separation, noxious stimuli and sedation vary with developmental stage. Infants respond with crying and agitation but are sometimes consolable. Toddlers and school-aged children are more resistant to comforting measures, unable to suppress their sometimes exaggerated fears and anxiety, and may become combative. Adolescents may respond unpredictably depending on their level of emotional maturity, ability to adjust to loss of control and disinhibition during sedation.

- Upper gastrointestinal tract
  - Craniofacial, esophageal, and tracheal malformations may be encountered making passage of the endoscope difficult or impossible and compromising the airway. These include pharyngeal and mandibular hypoplasia, cleft palate, esophageal stenosis, choanal stenosis, and tracheomalacia.
  - The esophageal lumen of the newborn is approximately 8-10 mm in diameter. The gastroesophageal junction (GEJ) is located 15-20 cm from the lips. Lumen diameter and distance to the GEJ increase proportionately with age.
  - The thin gastric wall allows partial visualization of the liver. The abdominal wall is easily transilluminated to localize scope position.
- Biliary and pancreatic ducts
  - The major papilla orifice in a neonate can be superficially canulated during ERCP with catheters tapered to 34 F. Deep selective canulation of normal biliary or pancreatic ducts is generally not possible in the neonate.
  - Age-related changes in normal common bile duct (CBD) diameter have been determined by transabdominal ultrasound. The CBD diameter should not exceed 1.6 mm in a child less than 1 year or 3.0 mm in a young adolescent. Beyond this age, adult standards apply. The pancreatic duct should not exceed 12 mm diameter in a child.
- Small bowel. Except in large adolescents, examination of the small intestine usually requires an operative approach (laparoscopy or laparotomy-assisted enteroscopy).
- Colon
  - Children often have prominent lymphoid nodularity of the colon and distal ileum.
  - The haustral pattern of the colon is less well developed in young infants.
  - The ileocecal valve and distal ileum can be intubated in nearly any age or size child given a sufficiently small caliber endoscope. A pediatric gastroscope in the range of 5 to 8 mm diameter may be necessary for neonates.

Indications and Contraindications

General indications for upper GI endoscopy, colonoscopy, and ERCP are the same for adults and children. Specific pediatric indications are listed in Tables 26.1-26.3. Investigation of inflammatory, peptic, allergic, and congenital disorders predominates. Neoplastic and preneoplastic conditions are included but are relatively infrequent.

- Enteroscopy. Indications are the same as for adults, i.e., obscure source of bleeding or polyps. The incidence of obscure small bowel bleeding in children is lower than in adults.

Table 26.1 Indications for EGD in children

Common
GERD
Allergic gastroenteritis (especially cow milk protein allergy)
Peptic ulcer disease
Celiac disease
Helicobacter pylori gastritis
Foreign body ingestion
Caustic ingestion
Severe dysphagia (for PEG)
Crohn's disease
Uncommon
Infectious esophagitis (candida, HSV, CMV)
Anastomotic strictures (repaired esophageal atresia)
Peptic esophageal strictures
Achalasia
Varices
Vascular malformations
Graftversushost disease
Lymphoproliferative disease
Familial polyposis
Gastroduodenal dysmotility (manometry catheter placement)

- Endoscopic ultrasound. Indications are the same as for adults, i.e., tumor staging, characterization of submucosal lesions, assessment of varices and other vascular lesions, guided biopsies, and guided injections. Tumors appropriate for staging are very rare in children. Clinical utility has not been assessed for children.

Equipment

Each of the three major endoscopy equipment companies (Olympus, Pentax, and Fujinon) manufacture instruments suitable for children. The smallest diameter endoscopes have 2.0 mm diameter operating channels that limit use of certain accessories. (see Table 26.4 for age and size-appropriate endoscopes).

- Accessories. Nearly all endoscopic accessories utilized in adult patients can be used in children given appropriate indications. Some accessories are not available in sizes that can be used with small tip diameter endoscopes or advanced through narrow operating channels. Palliative stents for obstructing cancers are rarely if ever used in children.
  - Hemostasis

    Choose an endoscope with the largest tip diameter and operating channel that can be handled safely so that light transmission and suctioning are optimal and a full range of accessories can be utilized.
    - Injection catheters and electrocautery polypectomy snares are available in 5 F sizes that will fit through a 2.0 mm operating channel of the smallest pediatric endoscopes.
    - Multipolar electrocautery probes are available in 7 F sizes that will fit through a 2.8 mm operating channel of a midsize endoscope. Larger (10 F) cautery devices require the larger channel in a therapeutic endoscope.

Table 26.2. Indications for colonoscopy in children

Common
Allergic colitis
Ulcerative colitis
Crohn's ileocolitis
Juvenile polyps
Familial polyposis (e.g.,adenomatous, PeutzJeghers, juvenile)
Uncommon
Infectious colitis (*C. difficile*, CMV, HSV, TB)
Vascular malformation
Cancer surveillance
Lymphoproliferative disease
Graft-versus-host disease
Foreign body
Stricture
Colonic dysmotility (manometry catheter placement)
Sigmoid volvulus

  - Elastic band ligation adaptors are available for endoscopes with a distal tip diameter of 9.0 mm up to 13 mm. Overtubes provided with ligation kits should not be used in children.
  - Clipping devices and detachable loops require medium to large operating channels.
- Foreign body retrieval
  - Most simple grasping devices (e.g., baskets, rat tooth jaws, alligator jaws) are available in 5 Fr sizes that will fit the smallest operating channels. Some devices with larger or rotating jaws require medium to large operating channels.
  - Most overtubes are too large for children and risk tearing the esophagus or pharynx.
- Feeding catheters
  - Percutaneous endoscopic gastrostomy (PEG) catheters are available from many companies in sizes suitable for infants and children. These are designed for both "push" and "pull" techniques and are composed of silicone or polyurethane materials.
  - French sizes relate to the catheter shaft diameter. Internal bolster diameters are generally not listed but increase in size proportionately with the catheter shaft diameter. (Table 26.5).
  - Internal bolsters may traumatize the esophagus of newborn or small infants.
- Pancreatobiliary
  - The Olympus infant duodenoscope (PJF 7.5) has a 2.0 mm diameter operating channel permitting passage of 5 F or smaller accessories. A 4 F ERCP cannula works well for diagnostic contrast injection through this endoscope. A double lumen tapered tip papillatome aids selective filling of the bile duct. Papillatomes and retrieval baskets are available in 5 Fr size that can be used with the PJF 7.5 scope for stone removal in infants.

Table 26. 3. Indications for ERCP in children

Common
Neonatal cholestasis
Common bile duct stone
Sclerosing cholangitis
Choledochal cyst
Postoperative biliary stricture
Idiopathic pancreatitis
Pancreatic duct anomaly (e.g.,pancreas divisum, anomalous pancreatobiliary junction)
Pancreatic pseudocyst

Uncommon
Postoperative or posttraumatic bile leak
Malignant biliary or pancreatic duct obstruction
Intrapancreatic duplication
Pancreatic duct stone

- Most retrieval balloons, dilators, and stents require a larger operating channel (e.g.,≥ 3.2 mm) found in standard size diagnostic duodenoscopes, therefore, limiting their application in infants.

Technique

- The methods of preparing and sedating adults undergoing endoscopy must be modified for children due to differences in physiology and behavior.
  - Patient preparation
    - Explain the procedure in a manner that matches the child's level of intellectual and emotional development.
    - An advance visit to the procedure unit may diminish anticipatory anxiety.
    - Transition objects such as stuffed animals or blankets, familiar music, and the presence of a parent may also reduce anxiety during early stages of preparation and initiation of sedation.
    - Use of oral presedative (e.g., midazolam) prior to insertion of an intravenous catheter may reduce anxiety, combativeness, and unpleasant recall.
  - Bowel preparation
    - Cooperation with bowel preparation regimens in children is difficult and frequently a major stumbling block for successful colonoscopy. Supervised administration of oral cathartics and lavage solutions and enemas during an inpatient or prolonged outpatient stay may be necessary to ensure success. Preparation of infants is easier due to relatively accelerated intestinal transit times and predominantly liquid diets.
    - One of three regimens may be chosen depending on the age of the child and parental preference.
      - Oral Fleets phosphosoda preparation
        - Contraindicated for children under 6 years and children with renal insufficiency.
        - For age 6-12 years, give 60 cc total dose in two divided doses separated by 4 h and for age >12 years, give 90cc total dose in two divided doses separated by 4 h.

Table 26.4. Recommended endoscope tip diameters for children [1]

	Premature 1.5 to 2.0 kg	Term Newborn 2.5 to 4 kg	Infant Child 5 to 15 kg	Older Child 16 to 25 kg	Preadolescents to adults > 25 kg
Gastroscope	5 mm or bronchoscope	5 to 8 mm	8 to 9 mm	9 to 10 mm	10 to 13 mm
Colonoscope[2]	5 mm or bronchoscope	5 to 8 mm	8 to 10 mm	10 to 12 mm	11 to 15 mm
Duodenoscope		7 to 8 mm	8 to 9 mm	9 to 10 mm	10 to 13 mm
Enteroscope[3]				9 to 10 mm	10 to 11 mm
Echoendoscope[4]				9 to 10 mm	10 to 13 mm

Approximate maximal outer diameter of insertion tube tip. Exact diameters vary with manufacturer and model. A gastroscope may be substituted if a colonoscope of suitable diameter is not available. A gastroscope can be used for intraoperative enteroscopy in young infants and children. Limited use in small children due to large diameter and longer nonbending section in the tip. Ultrasound probes may be used in endoscopes with operating channels ≥ 2.8 mm diameter.

Table 26.5. Recommended PEG catheter sizes for children

	Premature 1.5-2.0 kg	Term Newborn 2.5-4.0 kg	Infant Child 5-15 kg	Older Child 16-25 kg	Preadolescent to Adult >25 kg
PEG catheter diameter	Consider surgical or percutaneous nonendoscopic	12 to 14 FR	14 to 18 Fr	18 to 20 Fr	20 to 24 FR

26

- Mix each dose with 4 ounces of water and give an additional 24 ounces of clear liquid after each dose.
- Clear liquid and enema preparation
  - Clear liquids only for 2 days prior to examination.
  - Children's Senokot liquid (nonalcohol) is given for each of two days prior to exam.

1-6 months	2.5 ml
6-12 months	5 ml
1-5 years	7.5 ml
5-10 years	10 ml
> 10 years	15 ml

  - A saline enema (10 cc/kg to max of 500 cc) is administered on the evening before and the morning of the procedure.
- oral PEG balanced electrolyte lavage solution (Golytely, Nulytely, Colyte)
  - A single oral dose of Senokot (see above) or bisacodyl (5-12 yrs: 5 mg; > 12 yrs: 10 mg) and a single oral dose of cisapride (0.3 mg/kg, max 10 mg) are given one hour before starting lavage solution.
  - Administer from 38 ounces (depending on age) of lavage solution orally every 10 to 15 minutes until passing clear or only lightly colored liquid stool.
  - May administer by continuous nasogastric feeding tube infusion if unable or unwilling to tolerate orally. Avoid rapid infusion rates in young infants to prevent acute abdominal distension and related respiratory compromise.

- Dietary Restriction. Presedation dietary restriction is necessary to minimize the potential for pulmonary aspiration of gastric contents, particularly acid. Children may be offered clear liquids (this includes breast milk but not other milk or formula) up to 2 to 3 hours prior to sedation. Infants less than 6 months may receive formula up to 4 to 6 hours before sedation. For patients older that 6 months, solids and nonclear liquids should be held for 6 to 8 hours before sedation.
- Sedation
  - Intravenous sedation can be used successfully for endoscopy in the majority of properly selected children. This is safe in most ASA class I and II patients, and carefully selected class III patients.
    - The depth of sedation required for children undergoing endoscopy often reaches deep sedation, a level of sedation in which the patient

maintains spontaneous respiration and protective reflexes but responds only to painful stimuli.

- A minimum of two nurse assistants are required to perform the proceduresafely in children: one supporting airway management and assessing vital signs, the other preparing or administering additional medication and assisting with biopsies and other endoscopic accessories as needed. A third assistant, not necessarily a nurse, may be needed to help restrain an agitated child.
- Equipment for monitoring vital signs, administering supplemental oxygen, and for resuscitation must be available for different age and size children.
- General anesthesia or deep sedation administered by an anesthesiologist is used for ASA class III, IV, and V patients or when additional factors are considered such as: 1) problematic anatomy or dysfunction of the oropharynx and upper airway, 2) especially painful procedures, 3) technically demanding procedures, 4) prior history of unsatisfactory sedation, and 5) major behavioral disorders.
- Medications
    - Similar medications may be used for both children and adults but dosages must be adjusted for size or weight (Table 26.6).
    - Topical pharyngeal application of lidocaine or ben zocaine may be administered in metered doses, continuous spray, or paste to reduce gagging.
    - Combined use of a benzodiazepine and a narcotic is ideal to achieve optimal anxiolysis, hypnosis, amnesia, and analgesia. Rapid onset, short acting derivatives—midazolam and fentanyl—are preferred to reduce total procedure and recovery times. Reversal agents flumazenil and naloxone are available for each respectively but are not recommended for routine use.

- Endoscopic technique

  Basic techniques resemble those used for adult patients.

  - **Esophagogastroduodenoscopy**
    - Biteblocks are unnecessary for edentulous infants. Infants and toddlers may be swaddled. Even older children may feel more secure when wrapped in some form of blanket restraint.
    - Direct inspection of arytenoids and vocal cords is usually possible, looking for evidence of acid reflux inflammation. Pharyngeal lymphoid nodules and tissue is normally prominent.
    - The location of the gastroesophageal junction, which varies with age, is recorded in centimeters from the incisors.
    - Loss of the normal vascular pattern in the esophageal mucosa is an early and relatively reliable sign of esophagitis. Biopsies should be obtained from the proximal, mid, and distal esophagus to help distinguish allergic or idiopathic diffuse esophagitis from predominantly distal peptic disease.
    - The angle leading to the pylorus is often acute in an infant or small child and requires moderate retroflexion of the endoscope within the stomach.

Table 26.6. Sedation medications for children

Midazolam:	Oral, rectal, or nasal presedation: 0.5 mg/kg (maximum dose 10 mg) IV initial dose 0.05 mg/kg, then titrate to maximum dose 0.3 mg/kg.
Fentanyl:	IV initial dose 0.5 to 1.0 mcg/kg, then titrate to maximum dose 5 mcg/kg.
Flumazenil:	IV 0.02 mg/kg (max 0.2 mg); repeat at one minute intervals
Naloxone:	IV 0.1 mg/kg (max 2 mg); repeat at 23 minute intervals to max 10 mg.

- The duodenal bulb is best examined upon initial entrance into the duodenum to avoid misinterpreting lesions that result from scope trauma rather than primary disease. Fine mucosal nodularity may be seen in the duodenal bulb often representing lymphoid hyperplasia or Brunner gland hyperplasia. The posteromedial wall may remain obscured from view in a small child or infant where the lumen is narrow and the angulation rather acute.
- Random biopsies should generally be obtained from duodenum, stomach, and esophagus even when no gross abnormalities are detected since visual inspection alone results in underreporting of histopathology and opportunities for repeating endoscopy in a child may be limited.
- Colonoscopy
  - Similar to adult colonoscopy, carefully titrated analgesia and frequent reduction of loop formation are key elements for pediatric colonoscopy. Maximal analgesia is needed during the initial stages while advancing the endoscope through the sigmoid colon and around the splenic flexure.
  - Either left lateral decubitus or supine positioning of the patient is acceptable and may be changed to facilitate easier passage of the endoscope.
  - Haustral folds are less well developed in young infants and tip position is more easily determined by transillumination through the abdominal wall. Other anatomic landmarks are similar to those of adults.
    - Lymphoid nodules are often seen scattered throughout the colon in children and nodularity is particularly prominent in the terminal ileum. Biopsies, however, are still appropriate since early polyposis can mimic the appearance of lymphoid nodules.
    - Given the difficulties with bowel preparation and sedation in children and the subjectivity of gross visual findings, mucosal biopsies should be obtained in nearly all cases to avoid repeated procedures.
    - Beyond simple biopsy, polypectomy is the most common intervention performed during pediatric colonoscopy and, therefore, an important skill to develop. Polypectomy is performed for both diagnostic and therapeutic purposes. Polyps should be removed once encountered, e.g., during scope insertion, since later identification may be unsuccessful.
    - Since inflammatory bowel disease is among the most common indications for colonoscopy in children, intubation of the terminal ileum is an essential skill for the pediatric endoscopist.

Outcome

- Comparative outcomes have not been carefully studied in pediatric endoscopy. Increasing utilization of endoscopy in children has led, in general, to improved diagnostic accuracy and reduced surgical interventions for problems such as bleeding, polyps, strictures, enteral nutrition access, common bile duct stones, and pancreatitis.

Complications

- The range and rate of endoscopic complications in children is less well studied but similar to those reported for adults.
- Morbidity and mortality associated with some complications in adults (e.g., post-ERCP pancreatitis) is lower in children possibly due to fewer and less serious underlying comorbidities. Recovery may be faster in children.
- The emotional trauma of recalling painful examinations and interventions must be considered.

Selected References

1. Wyllie R. Pediatric endoscopy. In: Sivak MV, ed. Gastrointestinal Endoscopy Clinics of North America. Vol 4. W. B. Saunders Company, 1994.
   A well-edited and heavily referenced review of pediatric endoscopy techniques for problems including abdominal pain, peptic disease, IBD, GERD, GI bleeding, foreign bodies, enteral nutrition, portal HTN, and hepatobiliary and pancreatic disease.
2. Fox VL. Upper Gastrointestinal endoscopy and colonoscopy. In: Walker WA, Durie PR, Hamilton JR et al, eds. Pediatric Gastrointestinal Disease. 2nd Ed. Philadelphia: Mosby Year Book, 1996:151-341.
   A basic review of GI endoscopy in children including a limited number of color illustrations. Framed within an encyclopedic reference text for pediatric gastrointestinal disease.
3. Ingego KR, Rayhorn NJ, Hecht RMet al. Sedation in children: Adequacy of two-hour fasting. J Pediatr 1997; 131:155-158.
   A critical analysis of the duration of presedation fast and its relationship to residual gastric fluid volume and pH in 285 children undergoing endoscopy.
4. Gilger MA. Conscious sedation for endoscopy in the pediatric patient. Gastroenterol Nurs 1993; 16(2):75-79
   A thoughtful discussion of one of the most challenging aspects of pediatric endoscopy.
5. Guelrud M, CarrLocke DL, Fox VL. ERCP in Pediatric Practice: Diagnosis and Treatment. Oxford: Isis Medical Media Ltd, 1997.
   A unique monograph loaded with fluoroscopic images and extensive references.
6. Liacouras CA, Mascarenhas M, Poon C et al. Placebo-controlled trial assessing the use of oral midazolam as a premedication to conscious sedation for pediatric endoscopy. Gastrointest Endosc 1998;47: 455-460.
   Optimal approach for reducing anxiety in children undergoing sedation for endoscopy.

Index

A

B

C

Index

D

E

F

G

H

I

J

L

M

N

Index

O

P

R

S

T

U

V

W

Y

Z